Mitosis and Apoptosis

Matters of life and death

I. D. Bowen, S. M. Bowen and A. H. Jones

School of Pure and Applied Biology
University of Wales
Cardiff
UK
and
School of Biomedical Sciences
University of Wales Institute
Cardiff
UK

CHAPMAN & HALL

London · Weinheim · New York · Tokyo · Melbourne · Madras

Published by Chapman & Hall, 2–6 Boundary Row, London SE1 8HN

Chapman & Hall, 2–6 Boundary Row, London SE1 8HN, UK

Chapman & Hall GmbH, Pappelallee 3, 69469 Weinheim, Germany

Chapman & Hall USA, 115 Fifth Avenue, New York, NY 10003, USA

Chapman & Hall Japan, ITP-Japan, Kyowa Building, 3F, 2-2-1 Hirakawacho, Chiyoda-ku, Tokyo 102, Japan

Chapman & Hall Australia, 102 Dodds Street, South Melbourne, Victoria 3205, Australia

Chapman & Hall India, R. Seshadri, 32 Second Main Road, CIT East, Madras 600 035, India

First edition 1998

© 1998 I.D. Bowen, S.M. Bowen and A.H. Jones

Typeset by Cambrian Typesetters, Frimley, Surrey, UK

Printed in Great Britain by TJ International, Padstow, Cornwall

ISBN 0 412 71070 6

A catalogue record for this book is available from the British Library

Library of Congress Cataloging-in-Publication Data available

Mitosis and Apoptosis

Contents

Preface

We hope that this book, with all its imperfections, will prove to be a timely review and comparison of the phenomena of mitosis and apoptosis. Research in both areas has burgeoned in the last decade and has revealed tantalizing similarities in the processes that drive cells to reproduce and to altruistically commit suicide.

The scope of the topic is infinite, since mitosis and apoptosis are involved not only in shaping metazoan life, both plant and animal, but also in its constant and homeostatic maintenance. Peturbations in this delicate balance between mitosis and apoptosis invariably lead to significant states of disease. Cancer and autoimmune disease are but two of major medical significance.

Recent advances in molecular biology have increasingly revealed the genetic basis of both mitosis and apoptosis. The latter phenomenon is emerging as a genetically programmed cell death in a developmental context, while both phenomena appear to be regulated by environmentally orchestrated oncogenes.

Although this book is aimed primarily at instructing, and hopefully inspiring, graduate and undergraduate students, the authors have attempted to annotate recent developments with a generous citation of references. This enables the book to be used as a potential research guide which may thus be of interest to research workers at MPhil and PhD levels and their mentors. It is hoped in the long run that apoptosis, like mitosis, will be taught in sixth-form colleges and schools.

The book is organized into six chapters. Chapter 1 introduces the reader to the concept of turnover in an attempt to define apoptosis and mitosis in a kinetic sense. Chapter 2 is a more conventional treatment of the structure and physiology of mitosis and apoptosis. Chapter 3 deals with the cell cycle and its molecular and genetic basis, while Chapter 4 deals with the emerging genetic basis of apoptosis. This chapter is divided into two parts, the first dealing with definitive suicide genes, the second with the increasing number of genes (often oncogenes) and gene products involved in and associated with apoptosis. The first part of Chapter 5 deals with signal transduction mechanisms and explains in detail the role of particular cytokine signals, receptors and death domains in promoting cell death. The second section deals with cells, such as natural killer cells and cytotoxic T lymphocytes, that react to these signals, as well as the role of various cytokines in protecting and initiating apoptosis. Chapter 6 is a conclusion and intends to draw threads together in the context of the kinetics, distribution, incidence and lastly therapeutic significance of apoptotic cell death.

I. D. B., S. M. B. and A. H. J., Cardiff, July 1997

Acknowledgements

The authors are particularly indebted to their collaborator Dr Herbert Morgan for producing amusing and biologically relevant cartoons and for his exceptional draftsmanship. Much of the graphic material presented in the figures was produced with the aid of Harvard Graphics™.

The authors also wish to acknowledge the considerable secretarial support given by Miss Lorraine McMullan and Mrs Janet Brennan, without whom the manuscript could not have been produced. They wish to thank Dr Malcolm J. Edwards for stimulating academic discussion and providing key information on the cell cycle. In addition, the authors are grateful to Mrs Susan Jones for help and encouragement throughout the preparation of the manuscript and to Mr Dewi Rhys Bowen of Ysgol Gyfun Gwynllyw for commenting on the suitability of the text for undergraduates.

Finally, Dr Geoff Bond assisted in chasing up lost references and pursuing literature searches, Carol James, Sean Duggan and Gareth Walters found and photocopied numerous references, and Mr Vyv Williams provided an efficient darkroom service.

CHAPTER 1

Matters of life and death

1.1 INTRODUCTION

'In the midst of life we are in death'; from day to day, season to season we are reminded of this strange paradox. Without death there can be no life as we know it. In an ecological context, life and death form essential parts of a natural cycle. This is chemically underlined by our understanding of such phenomena as the carbon and nitrogen cycles and the normal translocations of matter that occur in the environment. Matter from dead organisms is transferred into new life often in the form of food or nutrients. Thus, cells will die accidentally when multicellular organisms die, as part of the ecological cycle of life and death, and their contents will in due course be recycled.

At a cellular level, cells reproduce and are born usually by means of a process of cell division, or rather multiplication, called **mitosis** (Chapter 2). Under special circumstances, when it is necessary to form the germ cells, eggs or sperm, and to reduce the number of chromosomes to a single set, a somewhat different process called **meiosis** is employed (Chapter 2).

Normally, therefore, the growth of a multicellular organism is largely dependent on the process of mitosis to increase the number of cells. We shall be looking closely at the mechanisms of this process in Chapters 2, 3 and 5. Conversely, resorption of tissues during normal development or morphogenesis has been shown to involve a particular kind of physiological or programmed cell death called **apoptosis** (Chapters 2 and 4). Indeed, apoptosis is now being thought of as an equal and opposite force to mitosis and thus is clearly also involved in normal day-to-day tissue balance or **homeostasis**. Both processes should be thought of as normal components of the cell cycle. Some cells continuously renew themselves; these are called stem cells and occur in renewable tissues such as bone marrow, epithelia, sebaceous glands and gonads. Some tissues renew themselves relatively slowly, but can also leap into action when wounded, e.g. skin and liver. Other specialized cells seem to be permanently postmitotic and never divide after formation, e.g. nerve cells and the xylem of plants (Chapter 6). Interestingly, apoptosis occurs most commonly in mitotically active tissues and tumours, perhaps as a counterbalance, although importantly it has now been shown that apoptosis can also be induced in non-renewing tissues such as brain.

Before studying mitosis and apoptosis more closely, consideration is given to the broad context of these phenomena.

1.2 CHEMICAL AND CELLULAR TURNOVER WITHIN THE ORGANISM

Multicellular organisms may be viewed as specialized environments, maintaining their special form and function by drawing on materials from the surrounding ecological environment. Although the surrounding environment in an atomic sense contains the same kind of matter as the internal environment of the living organism, there will be important differences in the way in which this matter is organized.

Because the spatial relationship between organism and environment is essentially intimate, it is often difficult to predict where the external environment ends and where the organism begins. Indeed, the degree of intimacy involved reflects the dynamic exchange of matter between both. In cellular terms, the boundary between 'outside' and 'inside' will be marked by the presence of a plasma membrane. By its nature, however, this is a fluid boundary allowing selective elements and compounds to enter and others to leave. The plasma membrane is essentially leaky and is covered with minute transient pores, which can control the entry and exit of water and ions, and the rate of entry and egress is controlled closely by energized pumps. At a fine structural level, cells can and do eat and drink, i.e. phagocytose and pinocytose respectively. In this situation, small parts of the external world, including food particles or fluid, are surrounded, isolated and taken 'into' the cells as vacuoles or vesicles, phagosomes or pinosomes. Technically, therefore, phagosomes and pinosomes, although internalized, are part of the outside world. Even when these routinely fuse with internally generated digestive vacuoles or lysosomes, they form mixed digestive vacuoles or secondary lysosomes constituting the 'gut' of the cell, again functionally and structurally continuous with the outside world. Only after the interiorized matter is digested and broken down into its constituents is it selectively absorbed, to be built up and organized into the specific and particular matter of the organism.

Living cells and organisms are thus very selective in terms of the substituents and compounds they take up from the external environment. Very specialized mechanisms such as receptor-mediated uptake are often employed to select and lock in rare molecules from the environment. Some of these specially selected molecules may act as signals or hormones. Incidentally, viruses may dress up in coats that emulate such signals in order to get themselves delivered into cells. Viruses are subcellular parasites and we shall see later that they have evolved mechanisms to inhibit cell suicide or apoptosis so that they can subvert the cell's synthetic machinery to their own ends. From the virus's point of view it is obviously essential that infected cells are prevented from committing the self-cure of suicide, since without the cell the virus cannot survive, let alone replicate. After virus infection, the race is on between altruistic cell suicide and eventual catastrophic cell murder.

1.2.1 The organization of living matter

That the matter found in the biosphere is organized differently from that in the non-living world or geosphere may be simply illustrated by sampling both in terms of the frequency of occurrence of atoms, as illustrated in Table 1.1.

Oxygen is clearly currently the most common atom. Note, however, that in the non-living world carbon comes 11th while it has jumped to second in the living world. Hydrogen and nitrogen have also jumped from a low position in the geosphere to a higher position in the biosphere. These have become the elements of life. It is the same matter that is circulating, but some elements, especially carbon, oxygen, hydrogen and nitrogen, have been selected by living systems and the same matter has become organized differently from the way it was organized or rather disorganized in the non-living world.

Matter is not often selected in elemental form, of course, but rather as compounds, water, oxygen and foods such as carbohydrates, fats and proteins, very often from other organisms in a complex food chain or net, linking eventually via decomposer organisms with the non-living world. The same atoms, therefore, continually pass through and between organisms and between the living and the dead. If we labelled the atoms using radioactive isotopes we would quickly see that atoms come and go quite rapidly in all living things and link up with the dead. Atoms continually flow into living organisms and work their way through in a particular pattern of structure and function before flowing out again as waste matter and/or dead cells. Atoms in the form of compounds will take their place for a relatively short while in all living things, the transient patterns which they will take up being specifically dictated by the genes in each cell. The cells themselves also cycle in their own right. To this extent we are all materially very transient ghosts.

Strictly speaking, a living thing is not a finite piece of matter at all, rather a place where atoms come together for a time, taking their place for but a moment in a characteristic reproducible pattern of structure and function and then leaving to be replaced by others.

A living thing, then, is a dynamic system, which has been likened to a whirlpool or sink (Fig. 1.1).

Table 1.1   Frequency of occurrence of atoms in the geosphere and in the biosphere

Geosphere										
Frequency	1	2	3	4	5					11
Element	O	Si	Al	Fe	Ca					C
Biosphere										
Frequency	1	2	3	4	5	6	7	8	9	10	11
Element	O	C	H	N	Ca	P	S	Cl	K	Na	Mg

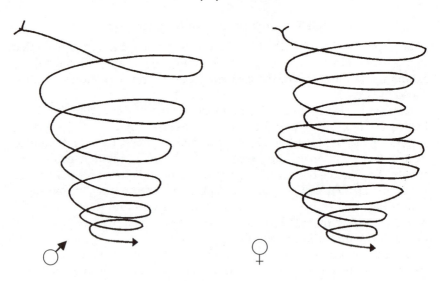

Fig. 1.1 Matter flows through the organism like a whirlpool, creating transient shapes in space (in this case a male ♂ and female ♀ torso). The matter is channelled through cells which have their own sinks and cycles.

These sinks do not operate in a homogenous context, but rather through a framework of cycling cells. In terms of the movement and turnover of matter, each cell continuously operates like a sink. The cells themselves also cycle and it is in this context that both mitosis and apoptosis play a crucial role in the shaping and maintenance of tissues and multicellular organisms.

In general, chemical turnover is more rapid than cellular turnover and some atoms may pass through the organism very rapidly, e.g. hydrogen, sodium and potassium. Others will linger longer; for example, calcium, which in humans literally hangs around in the skeleton. Thus, although materially and cellularly speaking we all have a different skin or gut from the one we had last week, we have more or less the same skeleton. It is in fact an extracellular product and is laid down as a matrix outside the bone-forming cells or osteoblasts. The exchange of calcium between the environment and the human skeleton is a vitally important process. Strontium, if present in the environment, will replace calcium in the bones. The 'dirty' atmospheric atomic bomb tests of the 1960s resulted in the release of radioactive strontium-90 into the atmosphere and its subsequent uptake into the long bones next to the blood-forming tissues of the bone marrow in children from high-rainfall areas of Wales, such as Aberystwyth. Unfortunately, strontium, like calcium, tends to linger for long periods in the bones, being laid down in a resilient extracellular matrix, and had an impact on the incidence of leukaemia in Wales.

There are plenty of examples to illustrate the dynamic nature of tissues and organisms in cellular and chemical terms. In a human context, the red blood cells have an average life span of about 120 days. It is estimated that overall almost 1% of all matter in the body is exchanged every day. Cells in the intestinal villus live only for a few days (Chapter 6). About 25% of blood serum proteins are exchanged each day and about 9% of all liver proteins. Indeed, all matter within the body will be exchanged in a very short while. Only a small percentage of matter and an even smaller percentage of cells remain continuously within the same living organism throughout its life, and most of that matter is usually tied up in extracellular matrices or skeletal material.

The material flux that occurs largely takes place via individual cells, but it is important to emphasize that most of these cells will be undergoing a flux of their own through the processes of mitosis/meiosis and apoptosis and that yet other cells will specialize and differentiate through a postmitotic phase to death.

It is incidentally fascinating that, despite all this chemical and cellular turnover, individuals remain more or less constant over a short period of time. Broadly speaking, each of us retains the same characteristics from day to day, or even from month to month. Your eyes and face, for example, will no doubt remain largely the same tomorrow and next week. This of course is the product of ongoing genetic activity. The cells constituting your eyes, nose, lips, etc. are receiving or have received specific molecular instruction to produce not only eye, nose and lip cells but **your** eye, nose and lip cells. This is specific instruction indeed.

How constant are these characteristics, however? Cells and organisms do change irreversibly in the longer term. People and organisms do mature, change and age irreversibly in time (Chapter 6). New gene activity, often environmentally triggered, results in the synthesis and appearance of new macromolecular species, leading to irreversible change or differentiation. If cells differentiate (**cytodifferentiation**) and specialize to such a degree that they cannot any longer divide, then death must be the ultimate consequence. Although this differentiation and death is genetically programmed it is not always identical (section 2.4) to what has been classically defined as apoptosis (Wyllie, 1981).

1.3 APOPTOSIS AND PROGRAMMED CELL DEATH

Individual cells within the organism may die either by accident or design. Accidental cell death may be caused by disease or other lethal external forces and leads to what pathologists call **necrosis** (Chapter 2). Death by design, often as part of a genetically regulated developmental programme, has been called **programmed cell death** (Lockshin and Williams, 1964; Lockshin, 1969, 1971). In this context, cell death is part of a strategy for survival of the organism. The paradox here lies in the fact that programmed cell death, more often than not, is essential for the continued life and development of the organism. Cell death is involved in the very beginnings of life. Thus, in the process of forming an egg the equivalent of three cells must die in order that meiosis can give rise to an appropriate surviving germ cell. The dead cells are discarded as 'polar bodies'.

It has for some time (Looss, 1889; Glucksmann, 1951; Saunders, 1966) been realized that cell death forms an essential part of the normal embryological development, morphogenesis and metamorphosis of life, where it forms an essential counterpoint to cell division or mitosis. Cell death in this context is employed to help shape and structure tissues, organs and organisms and can most spectacularly be seen in examples such as the transformation of caterpillar to butterfly, maggot to fly, the loss of the tadpole tail and the carving of the pentadactyl limb through interdigital cell death. These processes appear to be under genetic control, since the relevant mutants show specific disruptions of body plan (Chapters 4 and 6).

Less spectacularly, but equally importantly, programmed cell death plays an essential role in maintaining the population balance of cells in fully grown organisms. In terms of tissue kinetics such organisms are usually, more or less, in dynamic equilibrium. Cells are born and cells die by the hour. The overall shape and size of a tissue is normally maintained by this balance or homeostasis. A good example is the loss of dead cornified cells from the surface layers of the skin, which is essential for the production and maintenance of a protective skin layer. Cells are born through the process of mitosis at the basal granular layer of the skin and specialize in the production of a fibrous protein called keratin as they move up through the intermediate layers, finally differentiating to death as flattened, dead keratinocytes at the cornified surface. There is a similar renewal involving loss of cells in the villi of the intestine (Potten, 1992). A less well known example would be the maintenance of the size and shape of the liver, where it is estimated that a cell is born, and therefore one must die, every minute.

This homeostatic programmed cell death is an important phenomenon, and, indeed, led to a pioneering adoption of the term 'apoptosis', functionally defined as an equal and opposite force to mitosis (Kerr, 1971). It was surmised that a disturbance in this subtle balance between mitosis and apoptosis could lead, for example, to the formation of a tumour where cell proliferation outstripped cell death. Apoptosis was viewed as a positive type of programmed cell death with its own characteristic morphology, occurring not only during early development, but continuously throughout metazoan life. The word apoptosis itself was derived from the classical Greek word meaning 'a falling away', aptly used in the context of leaf fall, as we shall see, with its useful connotation of suicide, as opposed to the Greek word necrosis, 'to make dead', with its pathological connotation of cell murder (Duke, Ocjius and Young, 1996).

Since the endpoint of apoptosis and necrosis is the same, i.e. cell death, it is not always easy to distinguish between them, although more and more so-called diagnostic tools are being developed (Chapters 2 and 4). We may effectively be dealing with two defined extremes of one continuous spectrum of deaths (Wyllie, Duvall and Blow, 1984). From a definitive and comparative standpoint, then, apoptosis and necrosis are deemed to present the following characteristics.

Apoptosis always leads to cell condensation and shrinkage while necrosis leads to cellular swelling. Apoptotic cells tend to lose water while necrotic cells tend to gain water. Membrane pumps continue to function during apoptosis and apoptotic cells exclude vital dyes. Membrane pumps invariably fail early on in necrosis, leading to influx of water and sodium along with rapid calcium overload. Energy production is maintained during apoptosis; indeed, there is usually a synthetic surge leading to the appearance of new mRNA and protein species. ATP production drops catastrophically in necrosis as mitochondrial function is compromised by phospholipase A activation, a consequence of calcium overload, and there is an increasing acidosis leading to a 'salting-out' or precipitation of chromatin, resulting in a dark or **pyknotic** nucleus. In apoptosis, there is a dilation of the nuclear envelope accompanied by significant nuclear budding (blebbing), the blebs often being filled with chromatin, which characteristically moves to the nuclear margin. The DNA fragments under the influence of endonuclease activation, the enzyme cleaving the DNA at the internucleosomal linker regions. If run on a simple agar gel, such regular DNA fragments produce what is called a characteristic DNA ladder (Chapter 2). In the late stages, the lysosomes of necrotic cells burst and digestive enzymes are released which further lyse the cell as it breaks up (**karyolysis**). There is at this stage a complete loss of subcellular compartmentalization and the cell usually bursts, scattering its contents to such a degree that it elicits an inflammatory response. Apoptotic cells, on the other hand, exhibit a florid discrete fragmentation into physiologically intact spherical apoptotic bodies, which are immediately engulfed by near neighbours or macrophages without eliciting an inflammatory response.

Such a catalogue of differences would seem to be definitive; however, nature appears to be more complex and at this stage it would perhaps be appropriate to discuss a few contentious areas of definition.

It is possible for apoptotic cells to secondarily exhibit the properties of necrosis if they are exposed to a non-physiological environment. This is usually a condition associated with adverse *in vitro* conditions. Meanwhile, while apoptosis is normally induced by physiological stimuli, it has been found that chronic noxious stimulation of cells at 'subtoxic' levels can under certain circumstances induce them to commit suicide and become apoptotic (Cotter, Lennon and Martin, 1990). Increasingly, it has been shown that cells continually monitor internal levels of damage, especially DNA damage, and may decide in favour of apoptosis if such damage endangers the integrity and fidelity of future generations of cells (Wertz and Hanley 1996; see Chapters 4 and 5).

1.3.1 Apoptosis or programmed cell death?

There is considerable confusion in the literature between the terms 'programmed cell death' and 'apoptosis', and indeed other possible forms of cell death (Bowen, 1993). The term programmed cell death, coined in a developmental context by Lockshin (1969), has priority and on *a priori* grounds should have precedence in terms of generic usage. The situation is complicated by the fact that apoptosis as originally envisaged (Kerr, Wyllie and Currie, 1972) has a broadly morphological and kinetic definition while programmed cell death has a developmental and at least implied genetic basis. Apoptosis can be genetically driven, as seen by the classical work of Ellis, Yuan and Horovitz (1991) on the nematode *C. elegans* (Chapter 4), but some of the symptoms of apoptosis can apparently be induced without engaging in a genetic cascade (Duke, Sellins and Cohen, 1988). In other words, not all forms of apoptosis are genetically programmed.

It is also becoming increasingly clear that not all kinds of genetically programmed cell death are apoptotic in terms of its original definition (Bowen, 1993; Bowen, Mullarkey and Morgan, 1996; Lockshin and Zakeri, 1996; Szende *et al.*, 1995). Indeed, Clarke (1990) has detailed three basic morphological types of programmed cell death. One type of classification which would take most of these factors into account is shown in Fig. 1.2.

This classification envisages four types of cell death, three very similar to those described by Schweichel and Merker (1973) and one additional type, differentiation to death.

(a) Type 1 cell death

Type 1 cell death, as defined by Schweichel and Merker (1973) and Clarke (1990), accords with what we now call apoptosis, where there is clear cell shrinkage or condensation accompanied by chromatin condensation along the margins of the nucleus.

(b) Type 2 cell death

Type 2 cell death accords with that described by Clarke (1990) in neuronal cells and previously described by Lockshin and Williams (1964)

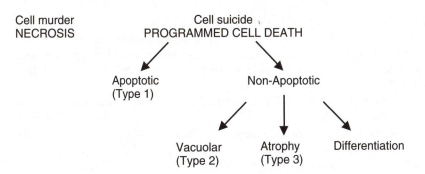

Cell murder   Cell suicide
NECROSIS   PROGRAMMED CELL DEATH

          Apoptotic   Non-Apoptotic
          (Type 1)

       Vacuolar   Atrophy   Differentiation
       (Type 2)   (Type 3)

Fig. 1.2 A classification of the different kinds of cell death into accidental, or necrosis, and planned, or programmed. The latter includes apoptotic (Type 1) and non-apoptotic types of cell suicide. Non-apoptotic cell death can be autophagic or vacuolar (Type 2), atrophic (Type 3) and also includes differentiation to death.

in insect metamorphosis as programmed cell death. This vacuolar or autophagic type of cell death was widely described (Bowen and Ryder, 1974, 1976; Bowen, Ryder and Dark, 1976; Bowen, Den Hollander and Lewis, 1982; Jones and Bowen, 1980) as **single cell deletion** in a range of invertebrates. It is characterized chiefly by the appearance of an increasing number of autophagic vacuoles leading to eventual cell fragmentation. Lockshin (1969) describes the dying midgut cells of a metamorphic insect as filling up with acid-phosphatase-positive digestive vacuoles before finally disrupting and releasing the still intact nucleus into the lumen. Bowen, Mullarkey and Morgan (1996) describe similar changes in detail in the dying salivary gland cells of the blowfly, where elevation in lysosomal and autophagic vacuole levels is followed by a *de novo* synthesis of ribosomal and cytosolic free acid phosphatase. Bowen and Lockshin (1981) clearly indicate that elevation of digestive hydrolases plays a significant role in final cell autolysis. Bowen and Bowen (1990) indicate that both lysosomal and characteristic free or extracisternal acid phosphatases play an important role in this type of cell death. An increased density of the nucleus has been described, sometimes leading to a clumping of chromatin but no distinct margination of the chromatin to the nuclear periphery, as seen in apoptosis.

(c) Type 3 cell death

Type 3 cell death is described by Clarke (1990) and Server and Mobley (1991) in neuronal cells and this kind of atrophic cell death (i.e. where the cell is deprived of an essential trophic factor) is described elsewhere (Martin and Johnson, 1991; Levi-Montalcini and Aloe, 1981) as occurring on the withdrawal of nerve growth factor (NGF). Dilation of the rough endoplasmic reticulum, Golgi apparatus and nuclear envelope occurs without dispersion of the ribosomes. The dilated endoplasmic reticulum eventually vacuolates but there is no overt autophagic involvement and nuclear changes are similar to those found in type 2 cell death (Fig. 1.3).

In an attempt to rationalize the situation, Lockshin and Zakeri (1996) have called apoptosis type 1 cell death and non-apoptotic programmed cell death type 2 cell death. Clarke type 3 cell death, considered here as akin to atrophy, as it is seen to occur in the case of nerve growth factor withdrawal, is likely to receive increasing attention since data is emerging to indicate that it can be induced in tumour cells by a range of kinase inhibitors (Szende *et al.*, 1995; Szegedi *et al.*, 1996; Mason *et al.*, personal communication; see Chapter 6).

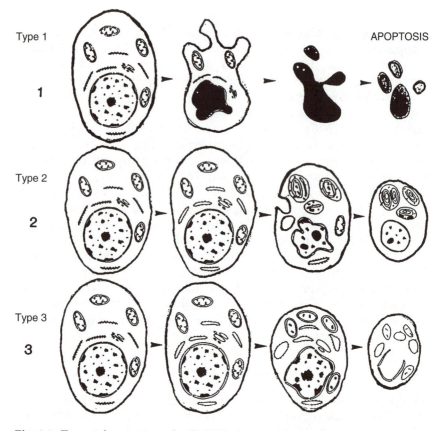

Fig. 1.3 Types of programmed cell death described in nerves: type 1, apoptotic; type 2, vacuolar or autophagic; and type 3, atrophic. Source: redrawn from Clarke, 1990.

1.3.2 Genetic control of cell death

Briefly, then, in relation to the classification presented here, necrosis displays characteristic symptoms of cellular swelling, gross calcium overload, phospholipase A activation and irreversible mitochondrial swelling leading to an autocatalytic lysis of the cell. It is never genetically controlled and, provided the induction is via a lethal stimulus, there is an immediate loss of energy production, water balance, volume control and protein synthesis. Necrosis appears to have no genetic control, while apoptosis can be genetically or non-genetically induced. Both types 2 and 3 cell death, as well as cytodifferentiation, appear to be genetically programmed and usually occur in a developmental context. Although these categories are largely based on morphological differences (Fig. 1.3) such as the presence or absence of chromatin margination and blebbing, such differences are also accompanied by differences in DNA fragmentation and protein synthesis (Server and Mobley, 1991).

Programmed cell death is usually dependent on genetic induction, although the inductive cascade (i.e. where a signal or regulator starts off a series of reactions, which can switch on a series of genes) can be accessed via activation of pre-existing transcription factors (Chapters 4 and 5). Good evidence exists for the genetic basis of apoptosis in the nematode *C. elegans* (Yuan and Horovitz, 1990) although some of the symptoms of apoptosis may vary in different species and can also be induced without engaging a genetic cascade (Chapter 5). Indeed, the current literature indicates that there are two broad types of apoptosis, one that is dependent on protein synthesis and thus amenable to genetic control and another that is independent of protein synthesis but dependent on the activation of pre-existing factors leading to kinase activation, which, as it were, prime the cell for apoptosis. This means that there are multiple pathways to apoptosis (Evans, 1993).

The morphological and biochemical bases of apoptosis have been well documented (Wyllie, 1980, 1981, 1987) and classically include cellular condensation accompanied by blebbing of the nuclear and plasma membrane. Nuclear changes include a characteristic margination of chromatin accompanied by endonuclease activation leading usually to internucleosomal DNA laddering. Evidence is also emerging in support of a genetic basis for non-apoptotic programmed cell death (Martin *et al.*, 1988; Martin and Johnson, 1991; Bowen, Morgan and Mullarkey, 1993; Bowen, Mullarkey and Morgan, 1996).

1.3.3 Differentiation to cell death

Finally, it is logical to regard cellular or cytodifferentiation as differentiation to death, since most differentiated cells are postmitotic and never regain the ability to reproduce and divide. To a large extent, therefore, differentiation, which is known to be based on differential gene activity, should be considered as leading to a special kind of programmed cell death. All differentiating cells move along specialized channels to death. Wangenheim (1987) concludes that differentiation into somatic cell lines makes cell death inevitable, in contrast with the potentially immortal undifferentiated stem- or germ-cell line. Some cells in certain organisms survive until the organism dies and others are continually produced in self-renewing tissues, differentiate to perform specific functions and then die.

1.4 MITOSIS

Prokaryotes and eukaryotes differ in the way they coordinate DNA synthesis and partition during cell division. In this book most of the data presented will refer to eukaryotic cells, where mitosis will be compared and contrasted with apoptosis. Mitosis occurs in diploid body cells (having two sets of chromosomes) to ensure the genetic identity of the daughter cells produced during cell division. Meiosis produces germ cells (sperm and ova) with a haploid (single) set of chromosomes ensuring genetic continuity between generations.

In prokaryotes DNA replication is followed immediately by cell division. In eukaryotes DNA synthesis and cell division occur in special phases of the cell cycle. Eukaryote cells can grow and divide at different rates; for instance, yeast cells divide approximately every 120 min. However, most plant and animal cells take 10–20 h to double in number. Some duplicate at slower rates and some differentiated 'postmitotic' cells do not divide at all, as indicated earlier.

DNA synthesis in eukaryotes does not occur throughout the division cycle but is restricted to a synthetic part before cell division or mitosis. A gap occurs in time after DNA synthesis and before cell division; another gap occurs after cell division, before the next round of DNA synthesis (Chapter 3). It is now understood that the cell cycle consists of an **M** (mitotic) **phase**, a G_1 **phase**, the first gap, the **S** (DNA synthesis) **phase**, a G_2 **phase** or second gap and so on back to M (Fig. 3.1). The phase between mitoses is called the **interphase**. Many postmitotic or non-dividing cells in tissues come out of the mitotic cycle into a resting phase called the G_0 **state**.

1.4.1 The phases of mitosis

Mitosis is the process that apportions the newly synthesized chromosomes equally between daughter cells. It has been recognized as a process for over a century and has been well described. During interphase between mitoses the chromosomes are not visible under the light microscope and the chromatin, consisting of DNA-protein complexes, is dispersed throughout the nucleus; then, during S phase, the DNA is replicated. This is followed by a characteristic series of microscopic events, which in fact unfold in a continuous manner. These events have been notionally subdivided into four substages: prophase, metaphase, anaphase and telophase (Fig. 1.4).

(a) Prophase

The chromosomes appear at the beginning of prophase as thin, stainable threads inside the nucleus. During the progression of prophase and metaphase they become visible as two identical coiled filaments, the **chromatids**, held together by a constricted region called the centromere. At late prophase the chromatids become shortened and more densely packed. Each chromatid contains one of the two new daughter DNA copies produced during the previous S phase. Therefore, each cell that enters mitosis has four copies of each chromosomal DNA strand.

Microtubules play a very important role in directing the subsequent movement of the chromatids. They form the scaffold of the

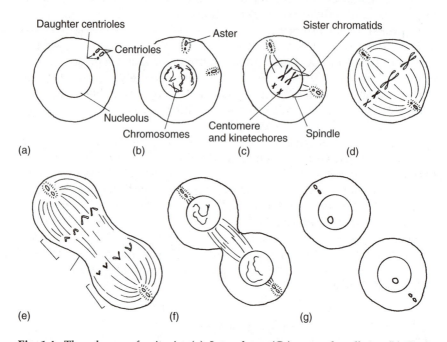

Fig. 1.4 The phases of mitosis. (a) Interphase (G_2) — each cell 4n. (b) Early prophase. (c) Middle and later prophase. (d) Metaphase. (e) Late anaphase. (f) Telophase. (g) Interphase (G_1) — each cell 2n. Source: reproduced from Lodish *et al.*, 1995 (from *Molecular Cell Biology*, 3rd edn, by Lodish, Baltimore, Berk, Zipursky, Matsudaira, Darnell). © 1995 by Scientific American Books. Used with permission of W. H. Freeman and Company.

mitotic cell and form tracks along which the chromatids move to ensure that they are apportioned equally to daughter nuclei. Cylindrical bodies called **centrioles** also play a key role in organizing this network of tubules. They are themselves microtubular in structure and replicate during interphase to form daughter centrioles. At the beginning of mitosis, two centrioles, each with its daughter centriole, move apart and microtubules radiate from each in all directions, forming **asters**. Centrioles eventually end up at opposite ends of the mitotic cell, now called **poles**. The radiating microtubules and associated proteins form a scaffold or spindle along which the separating chromatids will travel. Microtubules link the centrioles as they move apart and others connect with the chromatids via a granular region called the **kinetochore** at the centromeres of the chromatids; these catalyse the movement of the chromatids along the microtubules of the spindle. The end of prophase is marked by the disappearance of the nuclear membrane, and by metaphase the condensed sister chromatids can be found connected to the poles of the cell, attached to microtubules at their kinetochores.

(b) Metaphase

During metaphase the chromosomes become densely coiled and the sister chromatids all migrate to the equatorial plane of the cell, where they are oriented in the centre of the mitotic spindle.

(c) Anaphase

Anaphase is crucially marked by the simultaneous separation of all the sister chromatids at their centromeres. The kinetochores, now attached to separate chromatids, cause the sister chromatids or emergent chromosomes to migrate to opposite poles of the cell, ensuring that each new daughter cell receives an equal chromosome set.

(d) Telophase

Lastly, in telophase the chromosomes uncoil and the nuclear membrane for two daughter nuclei forms around the two sets of daughter chromosomes. At the same time there is a division or apportionment of cell cytoplasm and two cells separate where once there was one.

Plant cells, since they have rigid cell walls, show some variation from this basic plan and the progression of mitosis in yeast is somewhat more simplified: the daughter cell is formed by a budding process. Nevertheless, in all cases there is a quantitatively equal apportionment of genetic material. It should be noted that mitochondria and chloroplasts, where present, divide independently of the cell.

1.5 MEIOSIS

Meiosis is a form of cell division in which **haploid** germ cells (containing one copy of genetic material) are produced from **diploid** cells (containing two copies). This ensures that in sexual reproduction, when both haploid gametes fuse at fertilization, the full diploid complement of chromosomes and DNA is re-established in diploid zygotes. Generally, in meiosis, one round of chromosome and DNA replication, which makes the cell $4n$ (containing four copies), is followed by two separate cell divisions, rather than one, yielding four haploid (or $1n$) cells that contain only one chromosome of each homologous pair. The interphase between these two divisions uniquely lacks DNA synthesis.

CHAPTER 2

The structural and physiological basis of apoptosis and mitosis

2.1 THE RELATIONSHIP BETWEEN APOPTOSIS AND MITOSIS

Apoptotic cell death is seen in a large variety of metazoan cell types, especially those that have little cytoplasm (like thymocytes, lymphocytes or neutrophils) and are highly mitotic or have come from highly mitotic cell lines. In this respect, a certain amount of commonality exists between apoptosis and mitosis: indeed, both phenomena, being as it were equal and opposite in a kinetic sense, also share a number of specific common factors, which will be explored in the concluding section of this chapter. What is perhaps more interesting is that some common oncogenetic regulatory factors are also emerging. These will be touched upon here but are dealt with in greater detail in Chapter 4.

The relationship between mitosis and apoptosis is envisaged as a push/pull relationship akin to a seesaw, which may be balanced equally in a more or less dynamic equilibrium, may be biased in favour of growth, where mitosis exceeds apoptosis, or, conversely, biased in favour of resorption, where apoptosis exceeds mitosis (Fig. 2.1).

2.2 APOPTOSIS

The morphological changes themselves involve nucleus, cytoplasm and plasma membrane (Arends and Wyllie, 1991). The apoptotic cells round up, lose contact with neighbours and condense or shrink. The phenomenon was originally called 'shrinkage necrosis' (Kerr, 1971). During this shrinkage they lose surface features such as microvilli and desmosomes (Fig. 2.2).

In the cytoplasm, the endoplasmic reticulum including the outer nuclear envelope dilates. The cisternae of the reticulum swell to form vesicles and vacuoles, many of which fuse with the plasma membrane, giving the apoptotic cell a characteristic spongy appearance. The other cytoplasmic organelles remain intact; indeed, recent evidence indicates that the mitochondria can release activators that stimulate apoptosis.

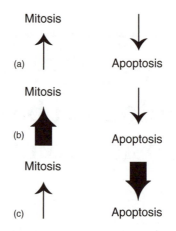

Fig. 2.1 The kinetic relationship between mitosis and apoptosis. (a) Tissues in a homeostatic dynamic equilibrium, where apoptosis more or less equals mitosis. Result: steady state. (b) Growing tissues or tumours where mitosis exceeds apoptosis. Result: growth and increase in size. (c) Resorbing, regressing or involuting tissues, where apoptosis exceeds mitosis. Result: regression and decrease in size.

There is good preservation of ultrastructure, although loss of water and shrinkage leads to a compaction of the cytoplasm, which results in an increase in cell density.

The plasma membrane of such cells becomes active and convoluted, eventually budding or 'blebbing' in such a way that the cell breaks up in a florid manner leading to a falling away (or, in Greek, *apoptosis*) of several membrane-bound spheres or **apoptotic bodies** of various sizes. Under physiological conditions these bodies remain viable and will exclude vital dyes such as trypan blue or nigrosine. They are not frequently seen in tissue sections, however, since they are rapidly phagocytosed by neighbouring cells or by macrophages, ending up within phagocytic and digestive vacuoles or lysosomes, where they are finally digested. This process of phagocytosis is usually so rapid that no inflammatory response occurs in the tissue.

Coordinated changes also occur in the nucleus of apoptotic cells. During the early phases, chromatin appears to condense and aggregates into dense compact masses along the margin of the nucleus, a process often referred to as **margination of chromatin**. The nucleus may become convoluted and may bud off into several fragments within the forming apoptotic bodies (Fig. 2.2). Where apoptotic cells or apoptotic bodies are shed into lumina they eventually lose membrane integrity and exhibit necrotic changes, often referred to as **secondary necrosis**.

Apoptosis is only one of several modes of cell death, as outlined in Chapter 1. Here it is compared and contrasted with necrosis and subsequently with mitosis.

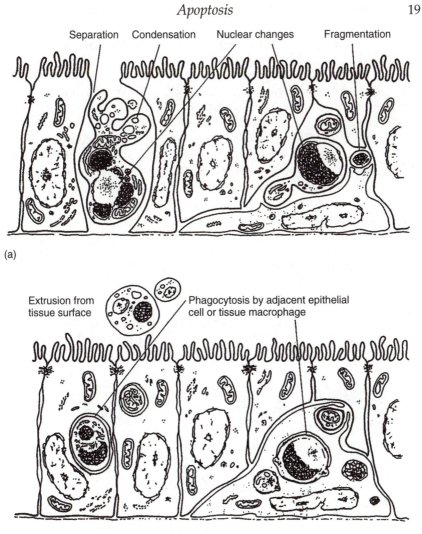

Separation Condensation Nuclear changes Fragmentation

(a)

Extrusion from tissue surface Phagocytosis by adjacent epithelial cell or tissue macrophage

(b)

Fig. 2.2 The morphological characteristics of apoptosis. (a) Formation of apoptotic cells. (b) Fate of apoptotic cells. Source: redrawn from Wyllie, 1981.

2.2.1 Nuclear events

One important characteristic of apoptosis appears to be DNA fragmentation. In the initial stages relatively large 50 kb and 300 kb fragments are produced, probably representing chromatin loops detached from the nuclear matrix. This is usually followed by a rapid double-strand fragmentation of DNA at the internucleosomal linker regions. This yields a series of oligonucleotide runs based on multiples of 180–200 bp

(Fig. 2.3), which can be separated out on an agarose gel as a regular 'DNA ladder'.

The enzymes that catalyse this fragmentation are usually non-lysosomal nuclear endonucleases. In certain cells they are activated by Ca^{2+} and Mg^{2+} and inhibited by Zn^{2+}. Other cells appear to have Ca^{2+}- and Mg^{2+}-independent endonucleases, and some invertebrate cells, such as those of *C. elegans*, have a lysosomal endonuclease that is only activated in phagocytosing host cells. Transcriptionally active DNA is preferentially hydrolysed by these enzymes and inactive regions are left as relatively long oligonucleotide fragments (Fig. 2.3).

Although endonuclease-induced DNA laddering was initially regarded as a diagnostic marker for apoptosis, it is now clear that not all cell types show this characteristic; indeed the nuclear changes of apoptosis can occur without endonuclease activation or oligonucleotide production (Oberhammer *et al.*, 1993).

2.2.2 Cytoplasmic events

The loss of water experienced by cells early on in apoptosis leads to an increase in density. Thus, apoptotic cells can be separated in a Percoll

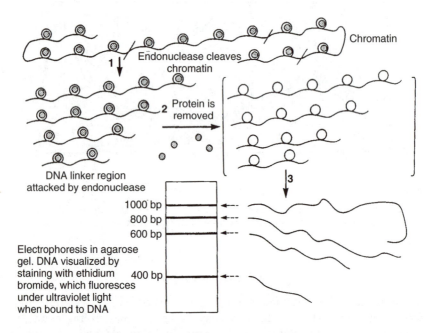

Fig. 2.3 A diagrammatic view of DNA laddering. (1) Euchromatin is hydrolysed by endonuclease. (2) The partially digested chromatin is deproteinized, giving fragments of DNA of varying lengths. (3) The fragments of DNA are separated by agarose gel electrophoresis.

density gradient, since they end up denser than normal cells. The cells also shrink and eventually break up to smaller, spherical apoptotic particles. These changes in cellular size and particle size can readily be measured by techniques such as flow cytometry, which can be used to separate off apoptotic cells.

All the mechanisms involved in the cell size and shape changes are not yet fully understood, but they do require considerable cytoskeletal reorganization. It is thought that depolymerization of actin (a contractile protein), activated by protein kinase C, is essential for the prolific budding process that produces apoptotic bodies. Some cells also exhibit activation of tissue transglutaminase, which appears to produce cross-linking of proteins in the cellular buds.

Several authors have reported increased transcription of specific mRNAs into apoptosis, although this seems to be followed later by activation of degradative RNAase activity, leading eventually to the rapid lysis of ribosomal and messenger RNA in the apoptotic cell (Chapter 6). There is still some controversy over the exact role of hydrolytic and degradative enzymes in the terminal phases of apoptosis. Some authors have presented evidence of *de novo* synthesis of hydrolases, such as acid phosphatase, collagenase and cathepsins, usually free in the cytoplasm and thus outside digestive vacuoles or lysosomes. Evidence is now emerging that specific proteases (e.g. interleukin 1β-converting enzyme; section 4.8.2) may even play an early regulative role in the process of apoptosis. This will be discussed in greater detail in Chapter 4, which deals with the genetic control of apoptosis, and Chapter 5, which deals with signal transduction.

2.2.3 Plasma membrane events

One important aspect of apoptosis involves phagocytosis of apoptotic fragments. This involves the recognition of apoptotic bodies by other cells and involves a number of plasma-membrane-based mechanisms (Savill *et al.*, 1993). One involves loss of sialic acid, thus exposing glycoprotein side-chain sugars. Another mechanism involves exposure and binding with phosphatidyl serine. A third mechanism involves secretion of thrombospondin by macrophages to form a molecular bridge between the apoptotic plasma membrane and the macrophage membrane (Fig. 2.4.).

2.3 THE MECHANISMS OF APOPTOSIS

Detailed aspects of the genetic mechanisms controlling apoptosis will be given in Chapters 4 and 5. Some general principles will be considered here. Apoptosis is seen as a characteristic and ordered series of morphological and biochemical events, although details may vary to some

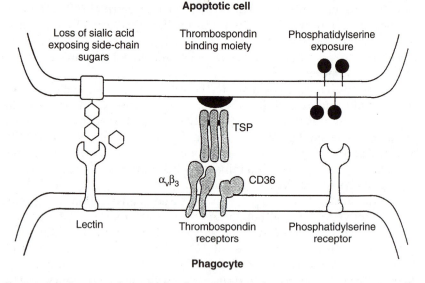

Fig. 2.4 Mechanisms by which phagocytes might recognize apoptotic cells. Source: reproduced from Savill *et al.*, 1993, with permission.

degree in different cell types. Thus, in the case of apoptosis in *C. elegans*, endonuclease activation occurs in the host phagocytic cell, not early on in the nucleus of the apoptotic cell as is the case with thymocytes. Nevertheless, there broadly appear to be two types of apoptosis: (1) unprimed and requiring new protein synthesis, and (2) already primed and not requiring new protein synthesis.

Genetically programmed apoptosis, in a developmental context, requires new protein and RNA synthesis, and inhibition of protein and RNA synthesis with agents such as actinomycin D and cycloheximide respectively will inhibit apoptosis in this context. Such inhibition can be seen in developmental examples of programmed cell death, as reported by Lockshin (1969), and also occurs in thymocyte apoptosis, as reported by Wyllie *et al.* (1984).

Many cases, including the induction of apoptosis in tumour cells by cytokines, indicate that effector molecules already exist in sensitive cells and protein synthesis is not necessary for apoptosis to progress (Chapter 5). Indeed, there are many examples now emerging where inhibition of protein synthesis itself triggers or at least enhances apoptosis, suggesting the possible existence of a short-lived protein inhibitor of cell death.

Apoptosis is therefore not an easy phenomenon to define in strict terms. The mechanisms that precipitate it have multiple pathways and a considerable level of redundancy seems to have been built into the system. This has led to the concept of states of readiness (or state of priming) for apoptosis (Arends and Wyllie, 1991), where the outcome is based on the metabolic induction or depletion of intrinsic effector molecules. The mechanisms of genetically programmed or 'cell autonomous' apoptosis are described in Chapter 4. The mechanisms of primed apoptosis are diverse but are probed in some detail in Chapter 5, dealing with cytokines and signal transduction.

2.4 NECROSIS

Necrosis is a pathological term used to describe the passive kind of cell death induced by lethal stimuli including trauma and disease. Necrotic cells characteristically swell rather than shrink. Plasma membrane damage leads to a loss of calcium, sodium and water balance, which is followed by acidosis and osmotic shock. The acidosis, or drop in pH, leads to the general precipitation of chromatin, leading to a darkened or 'pyknotic' nucleus. There is, however, no early margination of chromatin or activation of endonuclease activity, as usually seen in apoptosis. Trump, Berezesky and Osorino-Vargas (1981) stressed the importance of irreversible mitochondrial swelling in diagnosing necrosis. Both inner and outer compartments of the mitochondria become distended and dense deposits of matrix lipoprotein appear. Finally the endoplasmic reticulum and lysosomes swell and burst, the latter releasing digestive enzymes that induce further autolytic destruction of the cell, which finally breaks up, releasing inflammatory debris.

2.5 MITOSIS

Mitosis is an essential phase of the cell cycle, which leads to cell division. In mammalian cells, mitosis usually takes about an hour. During that period the cell builds up and then breaks down a specialized microtubular structure, the mitotic apparatus, a structure larger than the nucleus that is designed to capture the chromosomes (prophase), align them (metaphase) and finally separate them and daughter chromosomes (anaphase). These basic phases are illustrated in Figs 1.4 and 2.5.

The mitotic apparatus is organized into two parts, a central spindle of symmetrical microtubules and at each pole another set of tubules forming tuft-like **asters**. In each half of the spindle, a single **centrosome** organizes three distinct sets of microtubules. One set, the **astral microtubules** forming the aster, helps position the mitotic apparatus and will determine the eventual plane of cleavage when daughter cells separate during cytokinesis. Another set of microtubules constitute the **spindle**, and a separate set, the **kinetochore microtubules**, attach to the chromosomes at the kinetochores. Details vary in different species. The asters are absent in some organisms. In yeasts, where mitosis is simplified into a budding-like process, even the centrosome is absent.

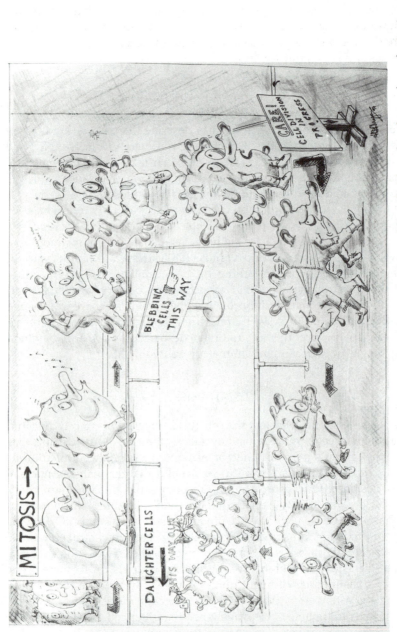

Fig. 2.5 A cartoon illustrating the continuous nature of mitosis (see Fig. 1.4 for internal detail). Note that the surface of mitotic cells becomes very irregular and develops blebs, like apoptotic cells.

2.5.1 The kinetochore

The kinetochores are specialized attachment sites at the chromosome centromeres. The centromeres are easily recognized as a constriction in the chromatin where sister chromatids are closely held and where the kinetochore microtubules are attached and link with a special segment of DNA, the **centromeric DNA**. The proteins comprising the kinetochores are currently being studied and several are known from their antibody-binding properties. One of the outer proteins is known to be dynein and similar work on yeast has shown the presence of proteins, such as the centromere-binding factors, that interact with the microtubules through other proteins, such as the kinesin-related protein Kar3p.

2.5.2 Duplication and migration

The assembly of the mitotic apparatus is initiated by centrosome duplication and migration during interphase and prophase. During prophase, kinesin-related proteins and cytoplasmic dynein participate in the movements of the kinetochores and chromosomal centromeres. During mitosis special 'motor' proteins called **kinesin-related proteins** (KRP) power the separation and migration of centromeres and attached chromosomes. A large number of KRPs have been identified in different roles from a diverse series of organisms ranging from yeasts to mice.

The formation of the mitotic apparatus and mitotic spindle involves the interaction of sets of dynamic microtubules. The microtubules are used to capture and slide the chromosomes via the kinetochores in a series of complex movements involving growth and elongation of transient microtubules followed by depolymerization and shortening.

At metaphase, forces generated at the kinetochores move the chromosomes to the equator of the mitotic spindle. The mechanisms of these complex movements are not fully understood but may involve changes in the flow of tubulin through the microtubules, involving rapid polymerization and depolymerization of the tubulin.

During anaphase, the daughter chromosomes separate and the mitotic spindle elongates. Initially, the kinetochore microtubules shorten and pull the chromosomes towards the poles. The two poles move further apart, pulling the separated chromosomes into what will become two daughter cells. In the later stages of anaphase, complex processes that involve sliding between polar microtubules, elongation of polar microtubules and forces exerted on the astral microtubules interact to sort out the separated chromosomes. Physiologically, it is known that these processes involve considerable ATP hydrolysis, kinesin-like and dynein proteins. Finally, the astral microtubules determine where cytokinesis takes place and the original cellular envelope cleaves to produce two daughter cells, each with a full set of chromosomes.

2.6 ELEMENTS COMMON TO APOPTOSIS AND MITOSIS

It has been suggested that apoptosis may be an abnormal or mis-timed mitosis (Ucker, 1991; Rubin, Philpott and Brooks, 1993). More specifically, Colombel *et al.* (1992) indicated that hormone-regulated apoptosis results from re-entry of differentiated prostate cells into an apparently defective cell cycle. Thus, the occurrence of mitotic death must not be ruled out. In a provocative paper, Raff *et al.* (1994) suggest that apoptosis ('programmed cell death') is orchestrated by a cytosolic regulator acting on multiple organelles in parallel, similarly to the way in which the cytosolic regulator M-phase-promoting factor orchestrates the mitotic phase of the cell cycle (Nurse, 1990). Such ideas evolve from the finding that apoptosis is frequently associated with disrupted growth factor production and reception, oncogenes and cell-cycle genes (Chapters 4 and 5). Indeed, it is probably still true to say that, in vertebrate systems, suicide genes as such remain to be discovered, whereas genes directly regulating an apoptotic programmed cell death are established in invertebrate systems — for example *ced-3*, *ced-4*, *ced-9* in the nematode *C. elegans* and the *reaper* gene in the fruit fly, *Drosophila*. Raff *et al.* (1994) further point out that apoptosis shares a number of features in common with mitosis; thus the cells are seen to round up, the plasma membrane becomes active and ruffles and buds, the nuclear lamina disassembles and the chromatin condenses. In both processes the subjacent nuclear lamina (which anchors the chromatin) is broken up by depolymerization of its constituent lamin filaments (Ucker *et al.*, 1992; Lazebnik *et al.*, 1993). Lamin phosphorylation and depolymerization occur during mitosis, catalysed by cdc2 kinase, and it is likely that the same mechanism operates during apoptosis (Nishioka and Welsh, 1994; Chapter 3). Genetic elements common to mitosis and apoptosis are discussed in detail in Chapter 4.

2.6.1 Apoptosis and the cell cycle

Where there has been investigation into the induction of apoptosis at different phases of the cell cycle (Sherwood and Schimke, 1994; Furuya *et al.*, 1994), using cell-cycle-phase specific drugs, it has been found that apoptosis can occur in any phase. These studies showed that the metabolic machinery necessary for apoptosis is present in the cell throughout the cell cycle. Indeed, apoptosis was found to be possible even after ablation of large parts of the total genome, suggesting that apoptosis effector molecules are constitutively present in the cell. The investigators also found that a short period of cell-cycle 'stasis' occurred before the onset of apoptosis and, although this waiting period varied in different cell lines, in general the duration of the pause was at least one cell-cycle length and usually of the order of about 18 h.

2.6.2 Apoptosis and cell-cycle genes

Elements of commonality between mitosis and apoptosis also extend to the involvement of oncogenes and cell-cycle genes (Chapters 3 and 4). Genes such as c-*myc* exhibit a dual role, the pathway triggered depending on associated growth factors and kinases. Thus, in the presence of the cytokine interleukin-2, upregulation of c-*myc* enhances lymphocytic proliferation, while upregulation of c-*myc* in the absence of interleukin-2 induces apoptosis. Levels of p53 produced by its tumour-suppressor gene may influence the survival of cells following DNA damage (Chapter 4). The induction of p53 provides the cell with an opportunity to repair damaged DNA before continuing to the synthetic or S-phase of the cell cycle. If the damage to its DNA is excessive, the increase in the concentration of p53 may direct the cell to self-destruct by apoptosis. Thus, tissues normally respond to excessive DNA damage by activating a p53-dependent apoptosis. It should be understood, however, that p53 is not critical for all apoptotic pathways (Chapter 4).

A number of inhibitors of the kinases that progress the cell cycle appear to induce apoptosis. For example, inhibitors of p34^{cdc2} (a serine-threonine kinase) lead to apoptosis and the gene *waf-1* (*p21*), an inhibitor of cyclin kinases, may be active in p53-mediated apoptosis; however, the role of p21 during p53-mediated apoptosis is controversial (Chapters 3 and 4).

CHAPTER 3

The cell cycle

3.1 INTRODUCTION

One of the key events in the life history of a cell is its division into two identical daughter cells, accomplished during a phase in its life known as mitosis (as described in Chapter 2). Another important period in the life of a cell, interphase, is, for many cell types, a preparation for mitosis. The two periods, mitosis and interphase, constitute the well-organized sequence of events known as the cell cycle. It is also important that events involving cell growth, DNA synthesis, chromosome segregation and cytokinesis are carried out in an orderly manner.

Controls exist within eukaryotic cell cycles of single cellular organisms, e.g. yeasts, so that once cells have replicated their DNA they do not do so again but proceed to mitosis, and cells that have not undergone mitosis cannot replicate their DNA (Nurse, 1994).

The story of how the cell cycle is regulated has begun to emerge in the last few years through cooperation between workers involved with the study of amphibian oocytes and yeast geneticists who have isolated yeast strains defective in some aspect of cell-cycle regulation. Information is steadily becoming available on the regulation of the cell cycle of higher eukaryotes and this could have far-reaching implications in the understanding of pathological transitions, especially those leading to uncontrolled cell proliferation (as seen in cancer).

The oscillation of cells between interphase and mitosis is controlled by what was previously known as maturation-promoting factor (MPF). This is now known to be a dimer made up of a protein known as a **cyclin** and a second protein that has kinase activity when bound to a cyclin. Two feedback loops are recognized for this MPF kinase activity. Firstly, MPF kinase activity is stimulated by MPF itself (an example of positive feedback) and, secondly, MPF stimulates the destruction of the cyclin by proteolysis, thus leading to a dramatic fall in MPF kinase activity (an example of negative feedback). Positive and negative feedback are key requirements for oscillatory systems to operate. The cyclic accumulation of proteins in interphase and their sudden destruction around mitosis provides the rational for naming these proteins 'cyclins'. The equilibrium between total (bound and free) cyclin concentration and that of active MPF is unstable, such that a small increase in cyclin concentration could lead to a possible tenfold increase in MPF activity.

Metazoans (multicellular organisms) possess stringent sets of checkpoints within the cell cycle that direct cell division. Additional checkpoints are required to prevent unnecessary and often disastrous overproduction of cells.

Development also takes control over the cell cycle. In embryos of, for instance, *Drosophila* and *Xenopus*, the progressive introduction of G_1 and G_2 phases of the cell cycle during development serves to slow down the early rapid alternations of DNA replication and mitosis that follow fertilization.

In response to DNA damage, metazoans have added a further complexity into the basic developmental programme that prevents cells from proliferating further but instead directs them to die by apoptosis.

3.2 THE STAGES OF THE CELL CYCLE

Biochemical studies on synchronized cultures of eukaryotic cells have shown that DNA synthesis in dividing cells is a discontinuous process and is separated from mitosis (M) by gap periods designated G_1 and G_2. This situation is almost universal in all higher organisms, including plants. Rather than considering G_1 and G_2 as periods during which little or no DNA synthesis is taking place, they may be alternatively considered as periods of general cell growth. G_1 represents a phase in which there is a great increase in the rate at which new components (except DNA) are made and G_2 represents a period during which a small amount of further growth takes place. Cells increase in size gradually

---

RATE AND ISOPYCNIC DENSITY GRADIENT SEPARATION OF CELLS

In a rate separation, particles (cells) become distributed in order of their sedimentation coefficients. In isopycnic separations, the particles (cells) become distributed within the gradient in order of their densities.

In a true rate separation, the maximum value of the density of the medium in the density gradient is smaller than the values of the density of the particles in the sample to be fractionated. Consequently, given enough centrifugation time, all particles will sediment to the bottom of the tube. However, before any of the particles are pelleted, centrifugation is terminated. Thus the particles are distributed through the density gradient as a series of zones, the positions of which are related to particle sedimentation rates.

throughout the cell cycle, making it possible to separate them by gradient centrifugation. Since cell density remains constant during most of the cell cycle, rate rather than isopycnic sedimentation is used in such experiments (see box on page 29)

Analysis of the total amount of DNA may then be done on cells isolated at different distances from the centrifuge rotor centre to demonstrate this discontinuous synthesis (Fig. 3.1).

The phase of DNA synthesis is designated the S phase and appears to be fairly constant in length in adult cells. The time cells spend in the S, G_2 and M phases varies surprisingly little: typically, in humans, they take about 12–18 h altogether. As cells become more differentiated they tend to increase the span of the cell cycle. This is largely attributed to a lengthening of the G_1 phase, which can last from a few hours to months or even years (Fig. 3.2).

The rate of incorporation of [^{3}H]-uridine into cells, indicative of RNA polymerization, increases steadily during the G_1 and S phase of the cell cycle, almost doubling in late S phase. This is followed by a reduction in uptake of [^{3}H]-uridine in G_2, which decreases even further prior to mitosis. Uptake is near zero during mitosis itself (Fig. 3.3).

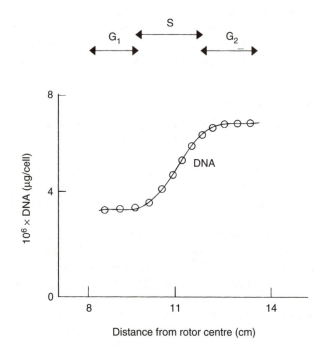

Fig. 3.1 DNA content of cells as a function of distance from the rotor centre.

DNA replication

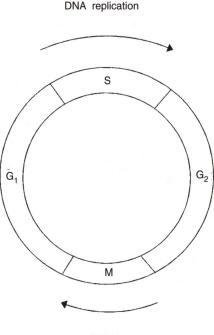

Mitosis

Fig. 3.2 The stages of the cell cycle. The cycle is made up of four phases: S (DNA synthesis), G_1 (gap 1), G_2 (gap 2) and M (mitosis). G_1, G_2 and S phases together constitute interphase

Caution must, however, be exercised in directly relating [^{3}H]-uridine incorporation with RNA synthesis because of possible turnover of RNA itself during the period of study.

The pattern of bulk protein synthesis in dividing cells may be demonstrated by the incorporation of e.g. [^{3}H]-proline into total protein, and parallels that of RNA, indicative of continuous synthesis throughout the cell cycle. Individual protein concentrations may well fluctuate during the phases of the cell cycle, indicative of specific requirements at certain times only. For example, histone proteins are manufactured largely during S phase when DNA replication is taking place.

It is, in part, the classical biochemical findings such as those described above that allow us to divide the cell cycle into the four phases, G_1 (gap 1), S (DNA synthesis), G_2 (gap 2) and M (mitosis) phase. The phases of the cell cycle are also defined by light microscopy in terms of changes in cell size and morphology. The most obvious and visually spectacular period of the cell cycle for microscopic characterization is mitosis itself.

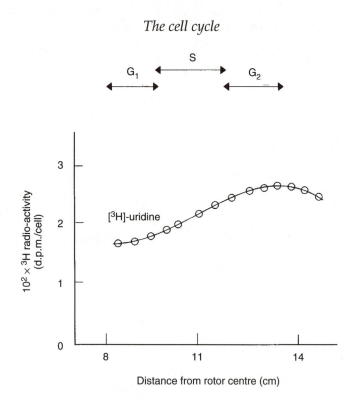

Fig. 3.3 RNA synthesis during the cell cycle measured by the incorporation of [³H]-uridine into cells at various stages of the cycle.

3.3 THE CELL CYCLE: A HISTORICAL PERSPECTIVE

Experiments conducted in the early 1970s led to the discovery that a factor was produced by mature eggs from the frog (*Xenopus laevis*) that, when microinjected into immature oocytes (arrested in G_2 phase of meiosis I), made them undergo meiosis and mature (arrested in metaphase of meiosis II), thus preparing them for fertilization. Normally, the hormone progesterone induces this maturation. This factor, which was clearly shown not to be progesterone, was given the name 'maturation-promoting factor' or MPF.

This 'frog oocyte assay' allowed extracts from a range of somatic eukaryotic cells to be tested for the presence of MPF. The success of the assay depended on the appearance, during maturation, of a white spot, which forms because the meiotic spindle apparatus migrates from the centre of the cell to a new position under the surface of the oocyte displacing pigment normally found there. Extracts from cell types ranging from yeast to humans all had MPF activity. These extracts did not, however, have MPF activity at all stages of the cell cycle. Extracts from cells in G_1 or S phase of the cell cycle seemed to lack it. MPF activity, therefore, appeared to fluctuate during the cell cycle, rising as cells entered mitosis (or meiosis) and then dropping sharply after the cells had divided. Thus, it seemed that all the familiar aspects of the cell cycle could be caused by the addition or withdrawal of MPF.

The second line of research involved a genetic approach. Yeast geneticists were playing their part by identifying a series of mutations that induced cell-cycle arrest at various points. One of these, the cdc2 mutation (cdc stands for 'cell division cycle'), was identified in the fission yeast (*Schizosaccharomyces pombe*) by Paul Nurse and his colleagues at Oxford (Nurse, 1990).

Fission yeast (*S. pombe*) is a yeast that reproduces by dividing in half with a typical eukaryotic cell cycle of G_1, S, G_2 and M phases but with no breakdown of the nuclear envelope. The budding yeast (*Saccharomyces cerevisiae*) is a yeast that reproduces by budding with no visible chromosome condensation during mitosis and with a nuclear envelope that remains intact.

The *cdc2* gene product has a role to play in two separate phases of the cell cycle: G_1 and G_2. It is now known that it is a kinase with a molecular weight of 34 000. This kinase is in consequence referred to as p34.

In the early 1980s, Hunt and Rosenthal were studying the control of protein synthesis in sea urchin eggs, trying to detect the synthesis of new proteins when the eggs were fertilized with sperm. Unlike work done by others on clam eggs, in which three new proteins appeared shortly after fertilization, Hunt and Rosenthal were unable to detect the synthesis of any new proteins when they fertilized the eggs with sperm, but, when they triggered development artificially, a new protein appeared within 10 min of fertilization. This protein appeared and disappeared with each egg cell division. The cyclic accumulation and disappearance of proteins during the embryonic cell cycle in clams and sea urchins led to these proteins being named 'cyclins' (Fig. 3.4).

Isolation and purification of MPF proved to be a long and difficult process but was historically accomplished by Lokka, Hayes and Maller of the University of Colorado Medical School in Denver. These workers found that MPF contains two proteins with molecular weights of 34 000 and 45 000 respectively. Antibodies raised to the yeast kinase also bound the 34 000 protein of MPF, thus showing that these two proteins were identical. Subsequent work has shown that the second component of MPF is cyclin. The work with marine invertebrates, frog oocytes and yeast genetics therefore converged when subsequent biochemical analysis revealed that cyclins and the *cdc2* gene product, p34 kinase, are the essential components of MPF.

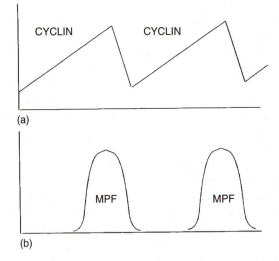

(a)

(b)

Fig. 3.4 The concentration of (a) cyclin increases until the kinase component of (b) maturation-promoting factor (MPF) is turned on at the midpoint of mitosis (metaphase).

3.4 YEAST AND HIGHER EUKARYOTE KINASES AND CYCLINS

The two yeast species *Saccharomyces cerevisiae* and *Schizosaccharomyces pombe* have become established as model organisms for the analysis of the cell cycle in eukaryotes because they show most of the characteristic properties of higher organisms.

In the budding yeast *S. cerevisiae* a single protein kinase subunit p34, the product of the *CDC28* gene, regulates progress through the G_1 checkpoint, known as 'start', the transition in the cell cycle that commits cells to enter S phase, and the G_2/M transition (Hartwell *et al.*, 1974). A single homologous kinase subunit, known as cdc2, that is very closely related to the *CDC28* gene product controls these key transitions in the fission yeast *Schizosaccharomyces pombe*. Thus p34[cdc2/CDC28] is a central molecule of cell-cycle control in yeast (Fig. 3.5). (Checkpoints have evolved to ensure that the start of a particular event in the cell cycle, e.g. DNA synthesis, is dependent on the successful completion of others, e.g. mitosis.)

The product of the *CDC2* gene, p34, is regarded as the prototype cyclin kinase subunit and as such serves as a yardstick for comparison with other cyclin kinases. The cdc2/CDC28 monomer itself has little or no kinase activity until it associates with a cyclin subunit. In yeast, different sets of cyclins cooperate with p34[cdc2/CDC28] at the different transition or checkpoints in the cell cycle. These cyclins disappear after completion of their function (Fig. 3.6).

In the G_1 phase of the yeast cell cycle, the cyclin Cln3, a 'non-cycling

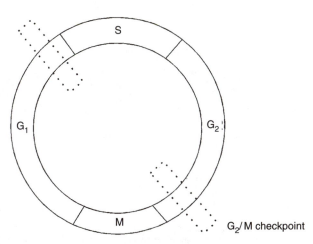

Fig. 3.5 The constitutive checkpoints in the cell cycle. One checkpoint exists at the G_1/S-phase boundary (known as 'start' in yeast) and a second checkpoint controls progression through the G_2/M-phase transition.

cyclin', promotes the accumulation of cyclins Cln1 and Cln2 and these associate with cdc2 (or CDC28) kinase to regulate transit through the G_1 checkpoint known as 'start' (Nasmyth, 1993). The cyclins Cln1 and Cln2 have nearly identical sizes (546 and 545 amino acids respectively) and are about 57% homologous. The genes for these three cyclins were identified by genetic experiments in *Saccharomyces cerevisiae* and, if all three genes are deleted, the cells are unable to go through 'start' and all die in the G_1 phase. Deletion of any one or two of the three Clns leads to an increased cell size and a prolonged G_1 phase.

During S phase, Cln1 and Cln2 are replaced by the cyclins Cln5 and Cln6 as p34 kinase's partners. At M phase, the mitotic cyclins Clb1 and Clb2 are the key partners for p34 kinase. Amino acid and DNA sequence analysis of the Clns and Clbs show that they are distantly related to one another. In mammalian cells, however, different cdc2-like kinases seem to separately regulate the critical cell-cycle checkpoints.

In experiments carried out on rat fibroblasts micro-injected with antibodies raised against p34, DNA synthesis occurred but cell division was prevented, thus showing that different kinases are involved at these two key points in the cell cycle (Ohtsubo and Roberts, 1993, 1995). The mammalian cdc2-related gene products are known as cyclin-dependent kinases or CDKs and include the p34^{cdc2} kinase. The presence of multiple types of CDKs in higher eukaryotes may reflect the increased regulation necessary to carry out the complex instructions that are required during development.

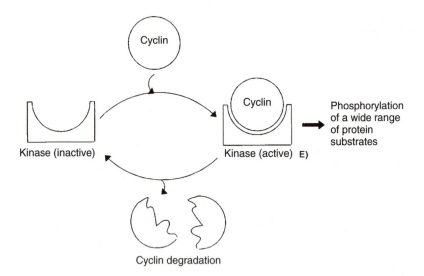

Fig. 3.6 As a cell grows, cyclin is synthesized and accumulates. This cyclin binds to and activates the kinase. The activated kinase, in turn, is able to phosphorylate a range of substrates. Eventually a protease degrades the cyclin and the kinase is rendered inactive.

A more detailed discussion of the various mammalian cyclins will follow later (section 3.5) but, for the present, the assignment of individual CDKs with their associated cyclins will be presented for the convenience of the reader.

The progression from G_1 to S phase involves cdk4/cyclinD and cdk2/cyclinE. Cdk2 is also implicated in S phase, when it becomes associated with cyclin A. The prototype cdc2 (p34), in association with cyclin B, is involved with the M phase of the cell cycle (Fig. 3.7).

From the above discussion it may be argued that cyclins D and E seem to be equivalent to the yeast cyclins Cln3 and Cln1/2 respectively, cyclin A is equated with Clb5 and Clb6 and cyclin B with Clb1 and Clb2.

It can be seen from the above discussion that several CDKs and cyclin subunits in higher eukaryotes undergo combinatory associations and regulate progression through the transition points in the cell cycle. It is important, therefore, to determine what distinct processes the different cyclin/CDK complexes control. It is attractive to speculate that the individual cyclins may be responsible for targeting their CDK partners to specific substrates at different points in the cell cycle.

3.5 HUMAN CYCLINS

There are, to date, five known types of human cyclin, designated A to E. Their classification is based largely upon sequence similarities and the

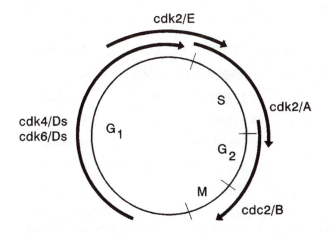

Fig. 3.7 Involvement of the various cyclin-dependent kinases (cdc2 and CDKs) and their associated cyclins during the mammalian cell cycle.

latter is generally confined to a 100–150 amino acid sequence, or 'cyclin box', as it is sometimes known, that is present in cyclins of all species (Poon and Hunt, 1992). This sequence of amino acids may be responsible for the interaction of cyclins with the CDKs.

Some human cyclins (cyclins A and B) have an amino acid sequence near their N-terminal ends, the 'destruction box', which is required for their destruction by the 'ubiquitin conjugation' pathway (see box in section 3.11). The destruction box might be recognized by a particular ubiquitin-conjugating enzyme. (Several different ubiquitin-conjugating enzymes are known.) A sequence rich in the amino acids glutamate, aspartate, serine, threonine and proline resides near the C terminal ends of some cyclins and this sequence is believed to contribute to the rapid turnover of the proteins.

Thus some features are present in most cyclins that mark them for either timing-triggered proteolysis by the ubiquitin pathway (Chapter 4) or for rapid turnover (Fig. 3.8).

Cyclin A protein is found only in the nucleus of cells and appears late in G_1 just before the start of DNA synthesis, slowly increasing in amount until the cell reaches prophase. It is degraded in metaphase by the ubiquitin degradation pathway. Cyclin A binds to p34 (cdc2) and to another kinase related to cdc2 that has a molecular mass of 33 000 and is known therefore as p33. Thus, it appears that cyclin A may be important in the transition from G_1 to the S phase of the cell cycle.

There are two forms of cyclin B in humans, B1 and B2. The former has been cloned. Cyclin B1 binds only to p34 (cdc2) and appears in the cytoplasm late in S phase and is then imported into the nucleus.

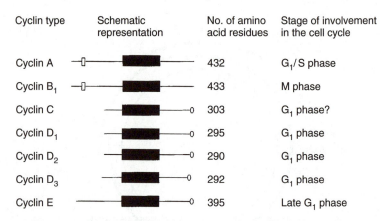

Cyclin type	Schematic representation	No. of amino acid residues	Stage of involvement in the cell cycle
Cyclin A		432	G_1/S phase
Cyclin B_1		433	M phase
Cyclin C		303	G_1 phase?
Cyclin D_1		295	G_1 phase
Cyclin D_2		290	G_1 phase
Cyclin D_3		292	G_1 phase
Cyclin E		395	Late G_1 phase

Fig. 3.8 Schematic representation of human cyclins. Filled rectangles represent the 'cyclin box', which is characteristic of cyclins. Open rectangles represent the 'destruction box' necessary for degradation by the ubiquitin pathway. Ovals represent sequences rich in proline, glutamic acid, serine, aspartic acid and threonine residues that contribute to rapid turnover of proteins.

Phosphorylation of the protein may be important for this importation. Cyclin B1 disappears sharply at metaphase, again by the ubiquitin pathway. As mentioned earlier in this section, cyclins A and B both possess the 'destruction box' necessary for this ubiquitin-dependent destruction.

Cyclin C possesses the PEST sequence near its C-terminal end and cyclin C mRNA has been detected in synchronized HeLa cells and appears to peak near the middle of the G_1 phase, well before cyclin A. Cyclin C is something of an enigma since no kinase partner or activity has so far been described.

The cyclin D members stand out because their levels do not oscillate. A similar situation exists for cln3 in yeasts. Cyclin D and cln3 may therefore be equivalent and may well act as initiator cyclins helping to coordinate cell growth with entry into the cell cycle. There are three D-type cyclins, D1, D2 and D3. The relative levels of cyclins D1, D2 and D3 vary between cell types. It is not clear whether they perform different roles in the cell cycle or analogous roles in different cell types. The retinoblastoma protein (pRb; see later) is a much better substrate for the D-cyclin-related kinases than the histone protein H1, thus suggesting that pRb may be an *in vivo* substrate for the D-type cyclins. One of the D cyclins, D1, has been identified as a candidate oncogene and is present in a region of chromosome 11 that has been amplified and overexpressed in 15–20% of human breast cancers.

Cyclin E protein begins to rise in mid G_1 and peaks near the G_1 and S-phase boundary. Cyclin E associates with cdk2. Thus cyclin E functions in G_1, preceding the role of cyclin A in the S phase of the cell cycle (Fig. 3.9).

Recently, a protein, p25, has been detected in brain tissue that associates

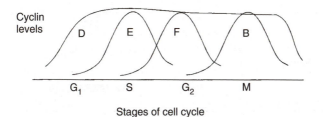

Stages of cell cycle

Fig. 3.9 The changes in the cyclin A, B, D and E pattern during the various phases of the cell cycle (G_1, S, G_2 and M).

with cdk5 and has a cyclin-like function. The active cdk5 kinase complex phosphorylates certain neurofilament proteins at sites identical to those phosphorylated by cdc2p34 and phosphorylates tau protein at identical sites to those phosphorylated in Alzheimer's disease.

Tau proteins are microtubule-associated proteins (MAPs) that participate in microtubule stability and the regulation of cellular architecture. A proportion of tau is highly phosphorylated in fetal and adult brain, whereas the majority of tau in the neurofibrillary tangles of Alzheimer's patients is hyperphosphorylated with many of the phosphorylation sites on seryl or threonyl residues.

Unlike cdc5, which is ubiquitously expressed in human tissue, p25 is expressed only in brain. The cdk5 kinase is the first example of such kinase activity in neurones and suggests that members of the cyclin-dependent kinases may be involved in processes other than cell-cycle control.

3.6 CELL-CYCLE CONTROL IN YEAST CELLS

Yeast cells possess controls that act during the cell cycle to ensure that S phase and mitosis are coordinated. Onset of S phase in the fission yeast (*S. pombe*) is brought about by the cdc10/res1 transcriptional complex, which activates the transcription of critical genes at 'start'. One of these genes is *cdc18*. The *rum1* gene product p25 controls the length of G_1 before 'start' (Peter and Heiskowitz, 1994). When *rum1* is deleted there is no G_1. Cells in G_2 are prevented from undergoing S phase because of the presence of the p34^{cdc2}/p56^{cdc13} complex. If cdc13 (a mitotic B-type cyclin) is deleted then cells undergo S phase repeatedly, generating high-ploidy nuclei. The p34^{cdc2}/p56^{cdc13} complex thus defines a cell in G_2 and

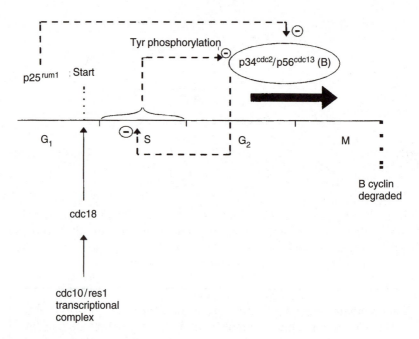

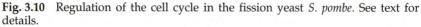

Fig. 3.10 Regulation of the cell cycle in the fission yeast *S. pombe*. See text for details.

when it is absent cells are able to reset to G_1 and undergo S phase (Moreno and Nurse, 1994; Fig. 3.10).

When budding yeast (*S. cerevisiae*) cells reach a critical size, they initiate bud formation, spindle pole body duplication and DNA replication almost simultaneously. All three events depend on activation of CDC28 protein kinase by the G_1 cyclins Cln1, 2 and 3. Activation of CDC28 by the G_1 cyclins (Clns) leads to activation of *cln1* and *cln2* transcription by a positive feedback loop. Activation of CDC28 by the Clns also inactivates the machinery responsible for the degradation of the Clbs or mitotic cyclins, thus allowing Clb levels to rise after G_1. Clb proteins then help to repress *cln* transcription while at the same time stimulating their own transcription in another positive feedback loop. The resulting increase in Clb levels leads to mitotic activation of CDC28, which then triggers Clb destruction. In wild-type cells, a protein p40[SIC1] appears at the end of mitosis and is a potent inhibitor of the Clb form of CDC28 but not of the Cln form of the kinase (Moreno and Nurse, 1994). The p40[SIC1] protein disappears shortly before S phase. Proteolysis of a cyclin-specific inhibitor of CDC28 is, therefore, an essential aspect of the G_1 to S phase transition (Fig. 3.11).

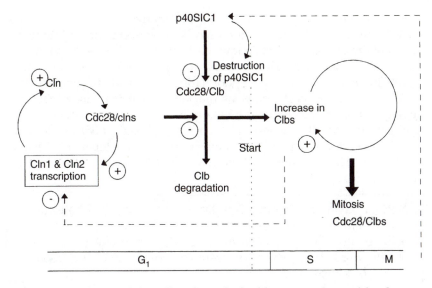

Fig. 3.11 Regulation of the cell cycle in the budding yeast *S. cerevisiae*. See text for details.

3.7 CYCLIN KINASE SUBSTRATES

In order to begin to appreciate some of the fundamental aspects of cell-cycle control, it is necessary to identify the various physiological substrates that are phosphorylated by the cyclin kinases that act at different stages of the cell cycle. This information, in turn, must be linked to key events that occur during growth and division of cells such as the replication of DNA, chromosome condensation and the breakdown of the nuclear membrane.

The above tasks are likely to be long and difficult ones for several reasons.

- While many potential substrates have amino acid sequence motifs (Ser–Pro/Thr–Pro) that make them candidate substrates for cyclin-dependent protein kinases, there are also a large number of other kinases with similar substrate specificities that may be responsible for their phosphorylation *in vivo*.
- Phosphorylation of candidate substrates may be an indirect process in the cell with the cyclin kinases acting only at the start of a kinase cascade.
- Some components that are phosphorylated either directly or indirectly by cyclin kinases may not be participants in cell-cycle progression.

With the above reservations in mind, a tentative picture of how CDK/cyclin complexes could control some of the fundamental processes of the cell cycle is presented below.

Proteins known as nuclear lamins polymerize to form part of the nuclear envelope. Upon entry into mitosis, profound structural reorganization of the cell occurs. Cyclin kinase (p34^{cdc2}/cycB) phosphorylation of lamin B2 at two sites triggers a cascade of reactions that ultimately results in another enzyme phosphorylating a third site (Courvalin *et al.*, 1992). These phosphorylations of the lamin proteins cause their disaggregation with the resultant break-up of the nuclear envelope. Also a lamin-associated 58 kDa integral membrane protein (p58) of the inner nuclear membrane is phosphorylated in a cell-cycle-dependent fashion by p34^{cdc2}. This phosphorylation is thought to cause dissociation of lamins from p58 (Chapter 2).

Nuclear lamins are intermediate filaments (IF) that are between 8 and 10 nm in diameter and classified on the basis of their amino-acid sequence as type IV IF proteins. They form highly organized sheets of filaments, which disassemble and reassemble at specific stages of mitosis.

Chromatin consists of 'spools' of histone proteins around which DNA is wound. Histones come in five basic forms designated H1, H2A, H2B, H3 and H4. The 'spools' consist of eight molecules — two of histone H3 combined with two of histones H4, H2A and H2B. The DNA is further locked in place by a single molecule of histone protein H1, which is bound to the outer surface of the DNA that wraps around the histone octamer. Histone H1 promotes the compacting of DNA in multicellular organisms. The histone protein H1 is extensively phosphorylated during chromosome condensation by cyclin kinase and such modifications might bring about the coiling and condensation of the chromosomes that occur as mitosis begins.

Other proteins such as high mobility group (HMG) proteins and several transcription factors are also substrates for cyclin-dependent kinases (Meijer, 1991). Phosphorylation of CDK sites on these proteins weakens their interactions with DNA. (The backbone of DNA carries a high density of negative charge at physiological pH. Phosphorylation of DNA-binding proteins would introduce negative charge to the latter, thus setting up adverse electrostatic interactions between these proteins and DNA.)

The intracellular proteins caldesmon and the regulatory light chain of myosin II are likely *in vivo* substrates for p34^{cdc2}/cycB (Yamakita, Yamashiro and Matsumura, 1992) and this fact may be of significance in view of the profound changes affecting the actomyosin system during contractile ring formation and cytokinesis. Phosphorylation of caldesmon causes it to dissociate from microfilaments and this process may facilitate microfilament disassembly at the start of mitosis.

RNA polymerase II is the enzyme that begins protein synthesis by transcribing genes into mRNA. Phosphorylation of the C-terminal domain of RNA polymerase II is thought to control the transition from transcription initiation to elongation. Many Ser–Pro/Thr–Pro motifs are found in this domain, which thus represents a possible site for multiple phosphorylations by p34^{cdc2} (Lu *et al.*, 1992).

The machinery responsible for chromosome segregation, the mitotic spindle, is regulated by cdc2 kinase. The molecular identity of the kinase substrate has recently been identified as an evolutionally conserved chromosomal protein of relative molecular mass 47 000 which has three sites for cdc2-kinase-mediated phosphorylation. The protein is phosphorylated only during mitosis. Mitotic arrest is induced by antisense mRNA or autoantibody, thus suggesting that this protein is required for progression through the cell cycle.

3.8 CYCLIN KINASE PHOSPHORYLATION AND DEPHOSPHORYLATION

Phosphorylation/dephosphorylation cycles that allow the interconversion of enzymes between active and inactive states or vice versa form a recurring theme in biochemistry. Examples range from enzymes that are instrumental in the regulation of intermediary metabolism, such as the mitochondrial pyruvate dehydrogenase complex, glycogen synthase, glycogen phosphorylase, to members of signalling systems such as the MAP kinase cascade.

The product of the *cdc2* gene (a kinase) and possibly other CDKs are not only controlled by protein–protein association (with a cyclin) but also by multiple phosphorylations. Binding of the kinase to a cyclin induces phosphorylation of the kinase at three sites. The *Wee1* gene codes for a kinase controlling the G_2 to M transition in *Schizosaccharomyces pombe*, and a *Wee1*-like gene that is structurally and functionally similar exists and is expressed in humans (Igarashi *et al.*, 1991). Phosphorylation at threonine 14 and tyrosine 15 by the Wee1 kinase is inhibitory, whereas phosphorylation of a conserved threonine 161 in cdc2 by a kinase known as the CDK-activating kinase (CAK) is required for activation (Fisher and Moran, 1994).

> Threonine 161 resides on a peptide loop that hides the substrate-binding site. Phosphorylation of this amino acid residue may induce a conformational change in the inhibitory loop and thus help stabilize the active conformation.

A CDK-related kinase has been identified as the catalytic subunit of CAK and a p37 regulatory subunit of CAK shares limited homology with members of the cyclin family, thus suggesting the existence of possible cyclin/kinase cascades. (The catalytic subunit of CAK has been renamed cdk7 and the p37 regulatory subunit cyclin H.)

The *cdc25* gene is a mitotic inducer controlling the G_2–M transition in *S. pombe* and a homologue of *cdc25* exists in humans (Kumagai and Dunphy, 1991; Strausfield *et al.*, 1991). A phosphatase, the product of this *cdc25* gene, removes phosphate groups from threonine 14 and tyrosine 15 and is thus activating. The activity of the cdc25 phosphatase increases at least fivefold at mitosis when cdc2 kinase activity is required for cell-cycle progress. Injection of antibodies to cdc25 blocks the cell cycle. The nuclear localization of cdc25 also emphasizes the accessibility of substrate to the enzyme.

It would appear that cdc25 is itself activated by phosphorylation: the active form has an extensively phosphorylated N-terminal domain, whereas the Wee1 tyrosine kinase becomes downregulated as a result of phosphorylation. The cdc25 phosphatase may be a direct or indirect substrate for cdc2 kinase itself. Thus, a positive feedback loop may exist generating a large quantity of active cdc2 kinase. In addition, cdc2 kinase may phosphorylate and activate itself, leading to a dramatic and auto-catalytic rise in its activity (Fig. 3.12).

There are three known CDK-activating phosphatases in humans, cdc25A, B and C. Cdc25A is expressed early in the G_1 phase of the cell cycle, cdc25B is expressed near to the G_1/S boundary and cdc25C is activated in G_2.

Recent work has shown that the product of the proto-oncogene c-*myc* forms a heterodimer with Max and that the Myc/Max heterodimer activates transcription of the *cdc25A* gene. The cdc25A phosphatase activates CDKs early in the G_1 phase of the cell cycle and, like c-*Myc*, can also induce apoptosis. The role of cdc25A in the induction of apoptosis is unclear.

3.9 DNA REPLICATION

It is instructive at this point to briefly put into place some of the key players involved at or near the replication fork of DNA, since subsequent sections of this chapter will refer to some of these components in the context of regulation of the cell cycle.

Structural distortion of the DNA double helix at the replication fork is brought about by the protein helicase, which interacts with the DNA. The helicase/DNA complex then binds the cellular replication protein A (RPA). This is a three-subunit single-stranded DNA-binding protein that allows more extensive unwinding of DNA by DNA helicase. Polymerase alpha and primase bind to the helicase RPA complex and an RNA/DNA co-polymer is formed with a base sequence complementary to the

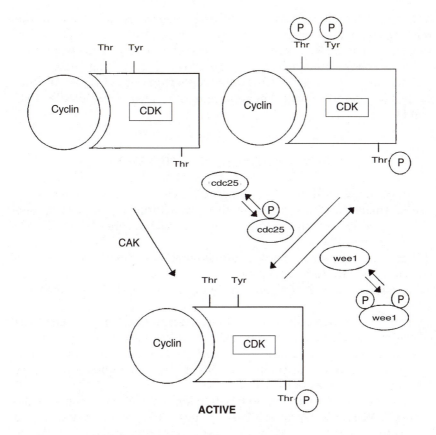

Fig. 3.12 CDK regulation by phosphorylation/dephosphorylation. CAK activates CDK by phosphorylation of threonine 161 whereas phosphorylation of threonine 14 and tyrosine 15 is inhibitory. Cdc25 phosphatase removes phosphate from threonine 14 and tyrosine 15 and is thus activating.

lagging strand. The DNA component of this RNA/DNA co-polymer is relatively short and is referred to as initiator DNA (iDNA). Polymerase alpha is thought to be required to form this iDNA and not the bulk replication of DNA. Initiation of replication of the leading strand is presumed to occur in a similar fashion. Note, however, that the leading strand is then replicated in a continuous manner, whereas the lagging strand is replicated discontinuously, generating 'Okazaki fragments'.

Replication factor C (RFC) displaces the polymerase alpha/primase complex and initiates the formation of a polymerase delta complex. This complex consists of RFC, polymerase delta and proliferating cell nuclear antigen (PCNA). This is the 'heavy duty' DNA replication complex. Note also that a polymerase alpha/polymerase delta switch has taken place. This complex then continues replication of the leading strand of DNA and the further elongation of the Okazaki fragments complementary to the lagging strand. Completion of replication of DNA is accomplished by the removal of the RNA primers by RNase H and MF1, followed by gap-filling and the eventual joining together of the 'Okazaki fragments'.

This brief overview of events at the replication fork of DNA puts into context key components of the replication machinery such as RPA, RFC and PCNA.

3.10 CYCLIN KINASE INHIBITORS (CKIs)

Much of the work done to elucidate cell-cycle control has involved the use of transformed cell lines. In these transformed cells, cyclins and CDKs associate in a binary complex that forms the core of the cell-cycle regulatory machinery.

Up to this time it has also been argued that these binary complexes are the primary effectors of cell-cycle progression. Further regulation is achieved by both activating and inhibitory phosphorylations of the CDK subunit.

In untransformed cells of, for instance, human fibroblasts, a major part of the cyclin-dependent kinases exist in quaternary complexes. These quaternary complexes consist of a cyclin, a CDK, PCNA and a fourth component. This fourth component has turned out to be an inhibitory cell-cycle control element or CKI.

One likely CKI candidate is p21 (section 3.12). Each member of the cyclin/CDK family can be inhibited by p21. This protein may therefore be an universal inhibitor of cyclin kinases. Ternary complexes can also be formed with cyclin/CDKs and p21. Therefore, it seems that PCNA is not necessary for interaction of p21 with CDKs.

As will be discussed in more detail later (section 3.12), p21 is a transcriptional target of the tumour suppressor protein p53, and p21 protein is absent in cells deficient in p53. On the other hand, overexpression of p21 may well inhibit cell proliferation in mammalian cells under certain circumstances (Fig. 3.13).

The CKI p27 is a potential inhibitor of G_1 cyclin–CDK complexes. In T cells leaving a quiescent state, interleukin 2 (IL-2) causes a decrease in p27 thus allowing CDK2 activation and entry into S phase. In contrast, IL-2 induces p21 and this p21 expression persists in cycling cells. It may be, therefore, that in normally cycling cells both p27 and p21 act together to establish a threshold that must be overcome for CDK activation to occur in the G_1 phase of the cell cycle.

It may be interpreted that p27 acts as a primary regulator of CDK activity as cells enter or leave the quiescent state, whereas p21 may be important in regulating CDK activity in cycling cells.

Another CKI candidate is p16, which interacts with the CDK4/

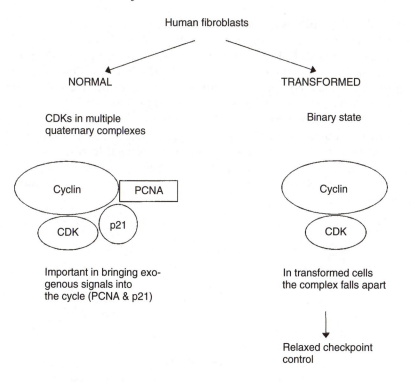

Fig. 3.13 Binary and quaternary complexes in transformed and normal cells. See text for further details.

cyclin D complex. CDK4 associates separately with p16, especially in cells that lack pRb. The protein p16 could possibly participate in a regulatory feedback circuit with CDK4, D-type cyclins and pRb. The relative abundance of p16 and D-type cyclins could determine the activity of CDK4 kinase and thus regulate cell-cycle progression. This negative feedback loop may be particularly important once Rb has been inactivated by phosphorylation. The *p16* gene is lost in a majority of tumour cell lines and in a considerable number of primary tumours.

> Most CKIs directly inhibit kinase activity by binding tightly to the CDK–cyclin complexes and, since CKIs are phosphorylated by their CDK targets, it is possible that CKIs interact with the kinase substrate binding site.

A new member of the p16 family, p15, has recently been isolated and its expression is induced 30-fold by treatment of human keratinocytes with transforming growth factor-beta (TGF-β). The gene encoding p15 is located adjacent to the *p16* gene on chromosome 9. CDK4 and CDK6 are both inhibited by p15, which suggests that p15 may act as an effector of extracellular growth inhibitory signals from TGF-β. Consistent with this view is the fact that TGF-β treatment causes accumulation of Rb in the underphosphorylated state. Withdrawal of cells from the cell cycle has been proposed as a component of differentiation, and induction of p15 by TGF-β could contribute to these processes (Fig. 3.14).

How can we make sense of these various cyclin-kinase inhibitor blocks during the normal operation of the cell cycle? Nasmyth and Hunt (1993) propose that 'cyclin kinase inhibitors may well have roles in the succession of cyclin waves during normal cell cycles'. They envisage that two constitutive CKI 'dams' exist in the cell cycle, the first operating in G_1 and the second in G_2.

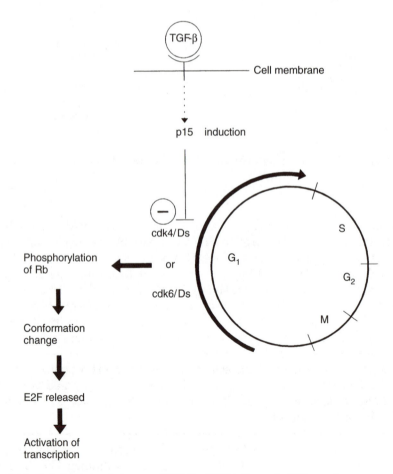

Fig. 3.14 Interactions between TGF-β, CDK4/6 and E2F, and movement through the G_1/S boundary.

In their 'hydraulic' model of cell-cycle progression 'CDKs build up like water behind two CKI dams'. 'Once these dams are full, the surplus cyclins trigger destruction or inactivation of the CKIs and the dams burst.' The cell is thus committed to S phase (the first constitutive block in G_1) and then on to the G_2/M transition (the constitutive block in G_2). Other inducible 'dams' may well exist in the cell cycle, built in response to DNA damage, starvation or growth factor limitation.

A subtly different view of the constitutive blocks in the cell cycle is that cyclin kinase activities are counterbalanced by the continuous presence of kinase inhibitors, such that 'spikes' of activity are generated when the kinase levels exceed an inhibitory threshold (Fig. 3.15).

Recent studies suggest that, in the case of one CKI at least (p21), multiple p21-containing kinases exist in both catalytically active and inactive states both *in vivo* and *in vitro* (Fig. 3.16; Zhang, Hannon and Beach, 1994).

The existence of these active quaternary complexes in normal cells may sensitize these cells to respond to changes in the growth conditions and external stimuli.

3.11 CELL-CYCLE ARREST IN OOCYTES

The oocyte presents an interesting problem in the understanding of cell-cycle control. Until it is penetrated by a sperm the cell cycle in the unfertilized egg is held in a state of 'suspended animation', locked in metaphase with chromosomes in a condensed state and aligned in the spindle ready to move apart.

What arrests the cell cycle in the unfertilized egg and how does fertilization trigger the continuation of the cycle? Cytostatic factor (CSF), recently identified as the product of the c-*mos* proto-oncogene, a threonine/serine kinase activity in the cytoplasm of unfertilized eggs, arrests cells in mitosis by preventing cyclin degradation. Chromatin and the cytoskeleton are maintained in a condensed and aligned state by a kinase

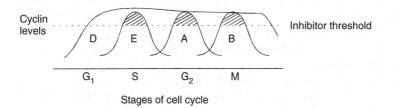

Fig. 3.15 Generation of 'spikes' of kinase levels (hatched areas) when cyclin concentrations exceed an inhibitory threshold. Letters A, B, D and E refer to changes in individual cyclin levels.

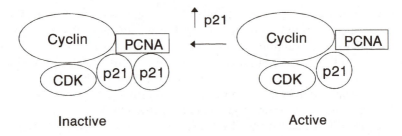

Fig. 3.16   Effects of changes in p21 levels and CDK activity. See text for explanation.

dependent on this cyclin for activity. In *Xenopus*, overexpression of c-Mos protein causes precocious meiotic maturation and prevents mitotic cleavage after fertilization.

At fertilization, membrane transduction signals bring about an abrupt increase in cytosolic calcium ions and a wave of increased calcium overwhelms the cell. This calcium activates a calcium-binding protein, calmodulin, which, in turn, is responsible for the activation of a calmodulin-dependent kinase, CaMK.

ATP-dependent phosphorylation of the proteasome, a hetero-oligomeric protease organelle of high relative molecular mass (see box), by this calmodulin-dependent kinase then occurs and cyclin is degraded.

Most proteins that are degraded in the cytosol are delivered to proteasomes. A proteasome is a non-lysosomal multicatalytic proteinase that is found in eukaryotic cells and is present in many copies throughout the cytoplasm. The proteasome has a 20S catalytic core (M_r = 700 000), which is responsible for ATP-dependent degradation of proteins that have become ubiquitinated. The 20S proteasome is composed of 14 subunits, which are distinct proteases. These proteases are arranged to form a barrel-shaped structure with their active sites facing the inside of the chamber. Large protein complexes formed from at least ten types of polypeptide form the floor and lid of the barrel and are thought to select the ubiquitinated proteins for destruction by binding to them and feeding them into the central chamber. The system of tagging proteins with ubiquitin — the ubiquitin ligase system — has three components: the ubiquitin activating enzyme (E1), ubiquitin carrier proteins (E2s), and a third enzyme (E3), which is responsible for catalysing the peptide bond between ubiquitin and the substrate protein. The enzyme E3 carries the protein binding site for the proteolytic substrate and therefore plays a central role in the selection of proteins for ubiquitin conjugation and subsequent degradation.

Cyclin is first bound to ubiquitin to mark it for destruction and interestingly the process of ubiquitination is itself controlled by phosphorylation. Degradation of CSF (c-Mos) is also triggered by calcium transients during fertilization but seems to take much longer than cyclin degradation.

The oocyte is very large — about 1 000 000 times the size of a normal cell — with a nucleus that is seemingly impotent to take care of the cell cycle. The newly fertilized egg may rely on its store of maternal RNA to produce a relatively simple cell-cycle control pattern in which fluctuations in cyclin levels determine when the cells divide. Indeed, early development is controlled by cyclins made under the direction of maternal mRNAs which are stored in the egg. Later on, after 12 initial rapid cycles of alternating mitosis and S phase, the embryo's genes begin to take over. The *string* gene in the fruit fly is one of the first embryonic genes to become active. When this happens the timing of mitosis is no longer regulated by the accumulation of cyclin but is regulated instead by the synthesis of the *string* gene product. Levels of *string* mRNAs remain at a constant level during the first 12 mitotic divisions and then begin to oscillate during subsequent cell divisioncycles, peaking in level during mitosis and falling again during interphase. The *string* gene is homologous to the *cdc25* gene in yeast and probably codes for a tyrosine/threonine phosphatase (see also Chapter 4).

Thus, we see that in developing embryos the initial rapid cycles may be slowed down by introduction of refinements such as G_1 and G_2 so that older embryos, even though cyclin and CDKs are still key players, have cell-cycle programmes that are run by developmental clocks (Fig. 3.17).

3.12 THE TUMOUR SUPPRESSOR P53 AND THE CELL CYCLE

Environmental mutagens and carcinogens are capable of damaging cellular DNA. In response to excessive DNA damage, human cells switch off DNA replication in order to avoid copying erroneous information. The tumour suppressor protein p53 is a key factor in this control mechanism and its expression is elevated under these circumstances. The *p53* gene is a tumour-suppressing gene that is not required for normal cell growth or division.

It is not understood how DNA damage generates a signal for increased p53 generation and stabilization. There are several enzymes within the cell nucleus that specifically bind to strand breaks in DNA and may be involved in generating the signal. These enzymes include poly-(ADP-ribose) polymerase (PARP), which synthesizes a nucleic-acid-like polymer from NAD^+ in response to DNA damage, three different DNA ligases that attempt to rejoin strand interruptions, and a variety of

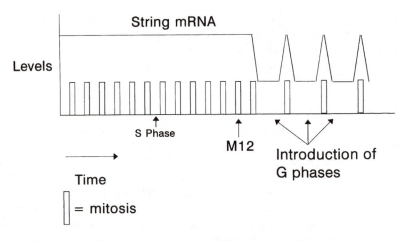

Fig. 3.17 Schematic representation of the early embryonic cell cycle. Alternations of S and M phases are associated with constant levels of string mRNA until the end of the 12th cell cycle when oscillations of string mRNA are associated with the progressive introduction of G_1 and G_2 phase.

DNA repair enzymes that recognize different forms of altered DNA. A large, DNA-dependent protein kinase is also activated by DNA strand breaks. Downstream events following p53 induction and stabilization are known to cause increased transcription of certain genes, including the gene coding for the recently discovered component designated p21. The p53 protein has a number of similarities with other proteins that regulate transcription. The N-terminus of p53 contains an acidic amino acid sequence characteristic of *trans*-activators, while the C-terminus has a helix–coil–helix motif with basic amino acids providing a DNA-binding domain.

The PCNA–DNA polymerase delta complex that is important in DNA replication is capable of binding the p53-induced p21 protein. When p21 is bound, the complex is inhibited. The cell thus has an opportunity to repair damaged DNA before continuing through the S phase of the cell cycle. Since PCNA–DNA polymerase delta is necessary for DNA replication, it is therefore necessary for p21 to be either inactivated or destroyed once DNA repair is complete, or there may be a selectivity in p21 function that allows it to arrest DNA replication while permitting active DNA repair (Li *et al.*, 1994). In support of this idea it would appear that p21 does not block PCNA-dependent nucleotide-excision repair.

If DNA damage occurs before the onset of S phase, then p21 inhibits cyclin E–CDK2 and so prevents cells from entering S phase. Inhibition of cyclin–CDK and inhibition of PCNA are independently executed by two different domains of p21. The N-terminal domain inhibits cyclin–CDK kinases and a C-terminal domain inhibits the PCNA–polymerase delta complex. Since p21 is able to inhibit cyclin–CDK complexes and the complex between PCNA and polymerase delta, it may provide a link between cell-cycle regulators and the DNA-replication machinery.

It is also possible that p53 binds directly to RPA. RPA is a multisubunit complex which binds to single-stranded DNA and this binding is one of the initial steps in DNA replication. Interaction of p53 with RPA inhibits the ability of RPA to bind single-stranded DNA.

The *RAD9* gene product and p53 may be seen to play analogous roles in yeast and mammalian cells respectively. In yeast cells, the *RAD9* gene is required to delay mitosis when chromosomes have been damaged by X-rays. Like p53 in mammals, the *RAD9* gene product is not required for normal cell growth or division in yeast but is expressed in response to DNA damage.

The induction of p53 in mammalian cells thus provides the cells with an opportunity to repair damaged DNA before continuing through the S phase of the cell cycle. If, on the other hand, the DNA damage is excessive, the elevation of p53 may direct the cells to self-destruct by apoptosis (Chapter 4).

3.13 THE TRANSCRIPTION FACTOR E2F AND THE CELL CYCLE

Several genes that code for proteins involved in DNA replication are activated in the boundary between G_1 and the S phase of the cell cycle. Many of these genes contain promoter regions that are capable of binding the cellular transcription factor E2F. E2F may therefore be a key regulator of the start of the S-phase of the cell cycle. The transcription factor E2F (molecular weight 60 000) is a DNA-binding protein that binds to DNA with the consensus sequence 5'-TTTSSCGC-3' (S = C or G). E2F has been cloned independently by different groups and represents a family of related proteins. To date, five members of this family have been isolated.

The DNA-binding domain of the E2F members carries the helix–turn–helix motif found in many DNA-binding proteins and is located near the N-terminus of the protein. It would appear that E2F family members form heterodimers (mixed pairs) with DP family proteins (transcription factors that help coordinate the cell cycle) and studies suggest that DP-1 (one of the family members) is hypophosphorylated during early cell-cycle progression and that its level of phosphorylation increases during progression through the cell cycle. This increase in phosphorylation is associated with a decrease in DNA binding of the DP-1/E2F heterodimer.

Gene products that may be regulated by E2F include dihydrofolate reductase, thymidine kinase, DNA polymerase alpha and thymidylate synthase. All these gene products are required for the synthesis of new DNA during S phase.

The level of *cdc2* transcription also increases at the G_1 and S-phase boundary and the *cdc2* promoter has an E2F consensus site. Deletion of this consensus site on the *cdc2* promoter results in an equal rate of transcription in serum-starved (non-cycling) and cycling cells, suggesting that E2F is important in regulating *cdc2* gene expression. Thus, it would appear that E2F plays an important role in the expression of genes at the G_1 and S-phase boundary.

Levels of E2F mRNA fluctuate during the cell cycle and increase at the G_1 and S-phase boundary. These fluctuations are consistent with E2F's possible role as a transcription factor at this stage of the cell cycle.

3.14 THE RETINOBLASTOMA PROTEIN AND THE CELL CYCLE

The retinoblastoma gene is a member of the tumour suppressor gene family and is only found in vertebrates. It is stable and constitutively expressed in most cell types regardless of their proliferative status. Inactivation of both copies of the gene in cells gives rise to eye tumours as well as other tumour types. The retinoblastoma protein (Rb) can exist in many states of phosphorylation ranging from the unphosphorylated to hyperphosphorylated forms. The phosphorylation state of the retinoblastoma protein changes during the various stages of the cell cycle. It is un- or underphosphorylated during most of the G_1 phase of the cell cycle (it is also underphosphorylated in G_0) but becomes phosphorylated on multiple sites at the G_1 and S-phase boundary (Fagan, Flint and Jones, 1994) and remains in this hyperphosphorylated state through the S and G_2 phase of the cell cycle. Up to 11 sites of phosphorylation (serine/threonine) have been proposed and the retinoblastoma protein can be phosphorylated on most of these sites by cdc2. It then becomes dephosphorylated during the M phase of the cycle.

Studies in yeast have shown that there is a point prior to S phase where the cell makes its decision to continue with the next round of cell division. This 'checkpoint' is known as 'start'. If yeast cells are exposed to mating pheromone prior to 'start', cell proliferation is inhibited. The 'start' checkpoint also prevents entry into the subsequent cycle if the cell has not reached sufficient size or has not accumulated sufficient macromolecules. Similar control mechanisms operate in animal cells. Withdrawal of growth factors before a checkpoint in G_1 prevents cells from moving on to S phase. The cell is then in a quiescent state referred to as G_0. It has, in effect, exited from the cell cycle. It is possible that Rb protein plays a key role at the G_1 checkpoint by helping to delay the onset of the S phase of the cell cycle until a suitable growth status is reached. The retinoblastoma protein may also allow cells to exit the cell cycle completely.

The term G_0 refers to a state of quiescence found in cell cultures but may be considered to be analogous to the terminally differentiated state. There is not a single, defined G_0 state. Quiescence represents a state outside the cell cycle and the G_0 state can have different depths.

An interesting property of the retinoblastoma protein is its ability to form complexes with the transforming proteins of several DNA tumour viruses by virtue of its so-called T-antigen-binding domain. The retinoblastoma protein is also capable of binding to E2F (Nevins, 1992). Additionally, the retinoblastoma protein has a DNA-binding domain near its C-terminus. The association of E2F with the retinoblastoma protein inhibits the transcription activator capacity of E2F but not the ability of E2F to bind at various promoter sites. Phosphorylation of the retinoblastoma protein by cyclin-dependent kinase, however, renders it unable to regulate the transcription factor E2F.

Factors that had been previously suppressed by the retinoblastoma protein can therefore then begin to be expressed. Two Rb-related proteins known as p107 and p130 also bind to E2F with similar functional consequences — i.e. transcriptional activity is suppressed. It would seem, however, that p130 binds to E2F during G_0 while p130 binds during the G_1/S transition. All three related proteins, Rb, p107 and p130, are collectively known as **pocket proteins** and seem to have characteristic temporal binding profiles with E2F (Fig. 3.18).

The retinoblastoma protein is able to interact with key cell-cycle components as well as with factors that are involved with the process of differentiation. Mice develop normally for about 13 days in the complete absence of Rb. This fact indicates that Rb is not required for the normal operation of the basic cell cycle. Goodrich and Lee (1993) have proposed that the retinoblastoma protein provides a link between higher-level regulatory networks and the basic cell-cycle clock.

As we progress from simple unicellular to multicellular organisms, the cell cycle itself becomes increasingly more complex, with additional layers of regulation. More complicated multicellular organisms are composed of highly differentiated tissues where growth factors, hormones, cell–cell contact and positional information are key determinants in the process of differentiation. Some of this information may be integrated and processed by a regulatory network and relayed to Rb. The retinoblastoma protein may therefore serve to link the higher regulatory levels of control with the basic cell cycle.

3.15 THE MITOGEN-ACTIVATED PROTEIN KINASE (MAP KINASE) CASCADE

Mammalian cells have a complete dependence on the presence of hormones to enable them to progress through the cell-cycle stages to S

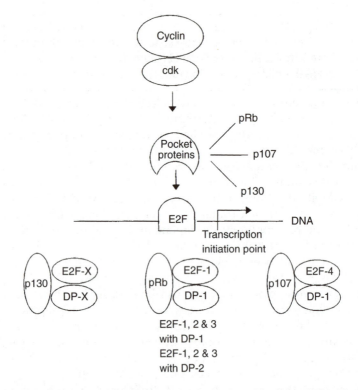

Fig. 3.18 Association of the pocket proteins (p130, pRb and p107) with the E2F family. More specific interaction is suggested between the pocket proteins and the E2F family/DP family heterodimers.

phase. Hormones that exert such functions are known as **growth factors** or **mitogens**.

When cells are stimulated by these mitogens a number of rapid changes occur. These changes transduce the signals produced by interaction of growth factors with membrane receptors into the interior of the cell. Many of these signals are sent to the nucleus via signalling cascades, resulting in a number of genes becoming transcriptionally active. In the final steps of these signalling cascades, specific transcription activators are phosphorylated and bind to sites in the control regions of the required genes to stimulate their transcription. Some genes may even be repressed rather than activated by these signals. Similar patterns of gene interaction are now emerging in mammalian apoptosis (Chapters 4 and 5).

The process of signal transduction is the acquisition and subsequent release of signalling information and has many features in common with the functions of electronic circuitry. There is considerable interest in these signalling pathways because if they go awry, cells may multiply uncontrollably, leading to cancers. **Proto-oncogenes** that regulate normal cell growth are key components of these signalling cascades. More than half of the known proto-oncogenes are directly involved in signal transduction processes. In mutated form (**oncogenes**) they may well contribute to oncogenesis.

One such proto-oncogene, *ras*, is central to a number of signalling cascades. Activating mutations of *ras* are found in nearly one-third of all cancers in humans. The *ras* gene products are members of a large family of proteins that act as molecular switches. Ras is in the 'on' state when it binds GTP. Hydrolysis of GTP to GDP leaves Ras in the 'off' state with this GDP firmly bound. Ras is itself a weak GTPase whose activity is enhanced by **G-activating proteins** or **GAPs**.

3.15.1 How is Raf turned on and off?

A large number of mitogens interact with membrane receptors in the plasma membrane of cells. Many of these membrane receptors are potential tyrosine kinases. Mitogen binding results in autophosphorylation of key tyrosine residues on the receptor protein. This phosphorylation facilitates interaction with controllers of Ras exchange factors such as Grb-2 (Sem5 in *C. elegans*). These, in turn, convert the inactive Ras-GDP to active Ras-GTP. This is accomplished by displacement of bound GDP with GTP.

The principal target of activated Ras is Raf, which is also a proto-oncogene product. Raf has an N-terminal domain that interacts with an effector region of activated Ras. This interaction localizes Raf to the plasma membrane. Ras then dissociates from Raf, leaving the latter in an active membrane-bound state.

Raf is a kinase (sometimes verbosely described as MAP kinase kinase kinase), which phosphorylates a second kinase known as MAP kinase kinase, or MEK, on serine and threonine residues. MEK is a kinase with dual specificity, capable of phosphorylation on serine/threonine and tyrosine residues. MAP kinase is phosphorylated by MEK and this covalent modification activates the enzyme and allows it to move into the nucleus. Among the targets of MAP kinase are proteins that release cells from cell-cycle arrest. These proteins become phosphorylated and as a consequence promote the transcription of genes associated with many different cellular processes. These genes include:

- cell-cycle regulators such as cdc2, CDKs and cyclins;
- DNA-binding proteins and transcription factors;
- proteins involved in signal transduction;
- enzymes involved in nucleotide and DNA synthesis;
- proteins involved in the building or degradation of the extracellular matrix, e.g. fibronectin, collagen and collagenase.

The Raf protein may also be activated by protein kinase C (PKC), which is stimulated by diacylglycerol released from membrane-bound phosphatidyl inositol bisphosphate (PIP_2) in response to mitogens that use this transduction route (Chapter 5). Thus different membrane transduction systems may be integrated by Raf into a common cascade.

A cell can thus receive a comprehensive picture of its environment and integrate the information through a variety of interconnecting intracellular signalling networks.

3.16 GENETIC ELEMENTS COMMON TO MITOSIS AND APOPTOSIS

From a genetic point of view, links operate between mitosis and apoptosis (Chapters 4 and 6). An arrest in the cell cycle may lead to apoptosis. The serine-threonine kinase $p34^{cdc2}$, which controls progression through the cell cycle when complexed with cyclins, has been observed to be rapidly activated after induction of apoptosis. Indeed, inhibition of $p34^{cdc2}$ results in absence of apoptosis. Thus, inhibitors of cell-cycle regulators may lead to apoptosis.

The protein produced by *waf-1* (p21) inhibits cdk2, a key factor in cell division (Douglas, 1994). p53-mediated G_1 arrest is dependent on induction of *waf-1*. This gene is an inhibitor of cyclin-dependent kinases that inhibits the Rb protein during G_1 and G_2. Rb is essential for progression of the cell cycle (section 4.7.8). p53 binds to two sites within the *waf-1* promoter. Waf-1 is a reversible inhibitor of cell-cycle progression that allows DNA repair. The exact role of *waf-1* in apoptosis is not fully understood but it is known that levels of Waf-1 increase as p53-mediated apoptosis occurs.

CHAPTER 4

The genetic basis of programmed cell death

A. SUICIDE GENES

4.1 INTRODUCTION

To date, the genetic basis of programmed cell death has only been defini-tively shown in a few cases and relates to developmentally induced cell death. These cases include genes controlling early neuronal development in the nematode *C. elegans*, and the *reaper* gene controlling embryonic development in the fruit fly, *Drosophila*. The type of cell death produced in both cases is apoptotic. Many other genes, however, have been identi-fied that have a bearing on apoptosis and these include mammalian proto-oncogenes, which also influence mitosis. This chapter will be broadly divided into two parts, the first dealing with apoptosis control genes in the nematode and fruit fly and the second half dealing with mammalian genes involved in apoptosis. This dichotomy is somewhat artificial, in that at least some homology is emerging between nematode cell-death genes and mammalian genes influencing cell death. Elements upstream of these primeval developmental genes are likely to be even more highly conserved and probably linked to homeotic master genes that dictate overall body plan in all multicellular organisms (see box on page 61).

In view of emerging data linking homeobox genes and proto-oncogenes with the control of both mitosis and apoptosis, a more comprehensive control network can now be usefully envisaged, at least in relation to developmentally induced 'endogenous' apoptosis (Fig. 4.2).

4.1.1 Plasticity of response

Despite the emergence of good evidence for the induction and inhibition of apoptosis by specific genes, it is clear that apoptosis independent of genes and of new protein synthesis can be induced by a wide range of epigenetic factors, which are not cell-lineage-specific. Such factors can also operate during normal development. Indeed, some workers claim

The hypothesis outlined here (Fig. 4.1) indicates that the genetic mechanisms that are cell-lineage-based and dictate body plan during early development in metazoa may be linked either directly or indirectly via gene regulatory proteins to cell-death genes and their products.

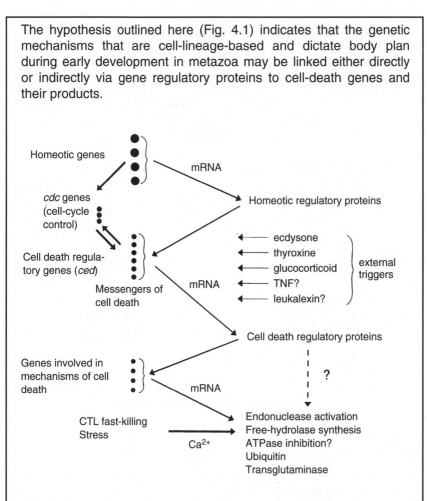

Fig. 4.1 A hypothetical composite of genetic mechanisms that may control the onset of programmed cell death. Source: redrawn from Bowen and Bowen, 1990.

Although this can be regarded as a 'hard-wired' version of apoptosis, it is not envisaged that such a cascade of genes operates on a simplistic one-to-one basis but rather as a multifunctional pleiotropic network.

that most mammalian cells normally express the proteins required to undergo apoptosis and that cell suicide is the default programme entered into if they are deprived of appropriate 'survival factors' from other cells.

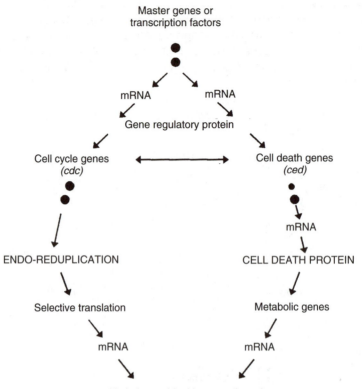

Fig. 4.2 A hypothetical basis for programmed cell death in insect development. The programme is envisaged as a cascading network of genes and gene products; homeotic master genes controlling cell death genes, which in turn influence cell cycle, endo-reduplication of DNA and metabolic genes leading to cell death. Source: redrawn from Bowen *et al.*, 1996.

4.2 GENETIC CONTROL OF APOPTOSIS IN THE NEMATODE *CAENORHABDITIS ELEGANS*

During the development of the nematode C. *elegans*, 131 of the 1090 cells generated undergo apoptosis. A total of 14 genes control this process and mutations in these genes exert specific effects on the cell death pathway (Fig. 4.3).

Two genes, *ced-3* and *ced-4* (*ced* stands for cell death) appear to induce apoptosis, while another, *ced-9*, inhibits apoptosis in cells destined to live. The products of *ced-3* and *ced-9* show some similarity to proteins that influence apoptosis in vertebrates (Hengartner and Horovitz, 1994a).

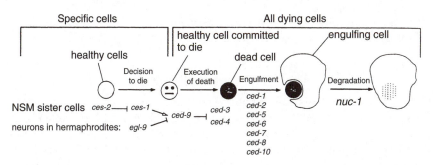

Fig. 4.3 The genetic basis of programmed cell death in *C. elegans*. The 14 genes illustrated here regulate cell death in the nematode's development. The initiation occurs in a few cells only, controlled by *ces-1*, *ces-2* and *egl-1*. Execution depends on the interaction between *ced-9* and *ced-3* and 4. All the other genes listed are involved in disposal and degradation of dead cells. Key: → = positive control; ⊣ = negative control. Source: reproduced from Hengartner and Horovitz, 1994c, with permission.

4.2.1 *ced-3* and *ced-4* are necessary for cells to undergo apoptosis

Mutations in either *ced-3* or *ced-4* prevent apoptosis occurring in the nematode (i.e. cell death abnormal), resulting in a cell-deathless animal which has an extra 131 cells. Genetic experiments show that both *ced-3* and *ced-4* encode 'suicide' functions within cells that die or within their close relatives (Horovitz, 1994). The *ced-3* gene encodes a 503 amino acid protein and is expressed during embryonic development. The protein is very similar to two mammalian proteins, namely interleukin-1β converting enzyme (called ICE) and the product of the *nedd-2* gene, an embryonic brain protein. ICE is a cysteine protease involved in cleaving 31 kDa IL-1β into the mature 17.5 kDa IL-1β. ICE protease is now also known to be involved in inducing apoptosis in mammalian cells (section 4.5). Among the ten or more members of the ICE protease family now known, the mammalian CPP32/Yamma/Apopain shows the greatest similarity with Ced-3 (Kuida *et al.*, 1996). Mice deficient for CPP32 showed abnormal development, particularly in the brain, where normal morphogenetic apoptosis, did not occur. Although the animals retained normal thymic susceptibility to apoptosis the normal pattern of cell death was not observed in the brain, where excessive numbers of cells persisted, leading to abnormal development.

The *ced-4* gene encodes a 63 kDa protein again expressed embryologically during programmed cell death. Recently, three functionally equivalent molecules to CED-4 have been found in mammals, and have been named apoptosis protease-activating factors (Apafs). Apaf-1 shows a remarkable similarity to CED-4, while Apaf-2, surprisingly, is cytochrome *c*. (Zou *et al.*, 1997). Upon binding to the cell death protease CED-3, CED-4 promotes apoptosis. Similarly, Apaf-1 could promote mammalian apoptosis by activating the mammalian cell death proteases such as caspase-3 (Chinnaiyan *et al.*, 1997a; Hengartner, 1997) (see section 4.8 for details on the cell death proteases).

4.2.2 *ced-9* inhibits apoptosis

Ced-9 actually controls the life and death decision of cells (Hengartner, Ellis and Horovitz, 1992). Thus, activity of *ced-9* can inhibit or negate the apoptotic-promoting effects of *ced-3* and *ced-4*. When *ced-9* is inactive then *ced-3* and *ced-4* activity results in apoptosis. The Ced-9 protein is similar to the product of *bcl-2*, a mammalian oncogene (see below): both genes appear to be members of the same family (Hengartner and Horovitz, 1994b).

Nematodes appear to share with mammals at least part of a common pathway to programmed cell death. Thus, overexpression of *ced-9* or *bcl-2* inhibits programmed cell death while overexpression of *ced-3* or ICE induces programmed cell death. Cloning and molecular characterization of *ced-9* has shown it to be an element of a larger locus also containing the gene *cyt-1*, which encodes a protein similar to cytochrome b560 in complex II of the mitochondrial respiratory chain. *ced-9* itself encodes a 280 amino acid protein showing sequence and structural similarities to mammalian proto-oncogene *bcl-2*. It has been shown that overexpression of *bcl-2* mimics the protective effect of Ced-9 on cell death in *C. elegans*.

4.2.3 Developmental control

How is the cell death programme regulated such that only the 'correct' cells enter into apoptosis during normal development in *C. elegans*? Master genes similar to those described in section 3.1. may operate upstream of the cell-death genes (Hengartner and Horovitz, 1994c). A few such genes have already been identified that specify cell death in a small number of specific cells, e.g. *ces-1* and *ces-2*. These master genes specify whether two specific motor neurone cells in the animal's pharynx will live or die. Ces-2 inhibits cell death and Ces-1 promotes cell death in these sister cells. Another gene, *egl-1*, can inhibit entry of two hermaphrodite-specific neurones into programmed cell death. The activities of these master genes appear to be lineage-specific and will influence the subsequent action of *ced-3* and *ced-4* in descendant cells (Fig. 4.3) and dictate the final outcome of life or death. In *C. elegans* these early events are followed by a sequence of activities controlling the recognition, engulfment and eventual digestion and disposal of the dying cells by phagocytic host cells. This chain of events is genetically controlled by seven genes in all (Fig. 4.3).

4.2.4 The engulfment and disposal of dying cells

As many as seven genes control the process of phagocytosis of apoptotic cells by neighbouring cells. Mutations in any one of the seven prevent engulfment of dying cells. The genes appear to fall into two groups, one causing major defects, *ced-1*, *ced-6*, *ced-7* or *ced-8*, and another group, *ced-2*, *ced-5* or *ced-10*, probably regulating the recognition of a dying cell by the phagocytosing neighbour.

One other gene, *nuc-1*, is required for the degradation of DNA. This gene is required not in the dying cell but in the phagocytosing host. Thus DNA in this case is not degraded by endonuclease in the apoptotic cell but rather by a DNAase activated in the lysosomes of phagocytosing host cells for the ultimate degradation and disposal of apoptotic cells. Mutation of *nuc-1* does not, therefore, prevent programmed cell death.

4.3 THE GENETIC CONTROL OF PROGRAMMED CELL DEATH AND DIVISION IN INSECTS

Insect development provides many good models for studying the role of both mitosis and programmed cell death in morphogenesis and meta-morphosis. Close inspection of insect models proves rewarding in that homeotic genes are known to be active in laying down the segmented body plan (Gehring, 1987) and some of these genes appear to control cell proliferation and may influence suicide genes, leading to the removal of larval tissues at the prepupal and pupal stages. In the latter case, large tracts of larval tissue may be deleted through programmed cell death responding to the hormonal cue provided by ecdysone. This large-scale controlled histolysis, which incidentally does not always follow the pattern of deletion set by classical apoptosis, is often followed by a more delicate pattern of apoptotic cell death in the imaginal disks (growing patches of embryonic stem cells) that go to form the emergent adult. Indeed, this suicidal 'fine-tuning' in the imaginal disks dictates the morphogenesis of the adult insect. The process is clearly genetically mediated, since different mutants display different patterns of imaginal cell death, which in turn dictates the phenotype of the adult.

Many connections are now emerging between genes active during *Drosophila* development and genes active in mammalian tissues and tumours. It has been shown that the *int-1* oncogene identified in mouse mammary tumours is similar to the *'wingless'* (*wl*) gene in *Drosophila* and that both genes influence cell proliferation. Gene products may act as morphogens or short-distance hormones, influencing the development of tissues nearby. Thus, the *Notch* gene of *Drosophila* is almost identical to the *lin-12* gene of *C. elegans* and the products of both genes enhance mito-sis, leading to massive neuronal proliferation in the *Drosophila* mutant 'big brain' and multiple genitalia in the nematode. Moreover, the peptides produced by these genes are homologous with EGF (epidermal growth factor), which occurs in vertebrates and mammals. The proliferative effects of EGF are thought to be mediated *via* a cell-membrane-bound receptor showing tyrosine kinase activity (as do *erbB* and other oncogenes). We also know that inhibition of tyrosine kinase activity can lead to programmed cell death (section 6.9.4).

There is good evidence supporting the genetic control of the pattern of mitosis in insect embryogenesis (Edgar and O'Farrell, 1989). Indeed, in *Drosophila*, the *stg* gene product responsible for cell-cycle arrest during early development has been shown to be homologous to cdc25, a regulator of mitotic initiation in yeast (section 3.8, including box). Such similarities may reflect common molecular mechanisms, since many of the genes involved in mitotic control are highly conserved (very similar) in yeast, flies, mice and men (Lee and Nurse, 1988). In addition, the characteristic modulation of specific patterns of mitosis and apoptosis early on during embryonic development is similar in a wide range of organisms, including marine invertebrates, nematodes, insects, amphibians, birds and mammals.

During *Drosophila* embryogenesis, mitotic control changes significantly during the 14th interphase. Before this, mitoses depend on maternal products and occur in waves. After this, zygotic transcription is required and mitoses occur asynchronously in a complex pattern. Mutations in the *string* (*stg*) locus cause cell-cycle arrest during G_2 of interphase 14, yet do not arrest other aspects of development. Stg is required to initiate mitosis and its predicted amino acid sequence is homologous to cdc25, a regulator of mitotic initiation in the yeast *S. pombe*. It appears that the regulated expression of *stg* mRNA controls the timing and location of embryonic cell divisions (Chapter 3).

4.3.1 Gene expression and gene products

Ubiquitin has been cited as a potential marker for impending cell death in insects (Schwartz, Kosz and Kay, 1990; Schwartz, 1991; Schwartz and Osborne, 1993). The type of cell death reported here in the tobacco hawkmoth, *Manduca sexta*, was more akin to Clarke type 2 rather than apoptosis, showing no DNA fragmentation and no nuclear and cellular blebbing (Schwartz and Osborne, 1993). Ubiquitin is a highly conserved 76 amino acid protein found in all eukaryotic cells and is thought to play a role in the removal and degradation of cellular proteins in subcellular proteasomes, including the cyclins involved in mitosis (see box in Chapter 3). Ubiquitin is added to lysine residues as a post-translational modification and this addition targets proteins for proteolysis via an ATP-dependent non-lysosomal proteinase. Although ubiquitin itself is unlikely to initiate cell death, it merits further study.

Selective gene expression has also been reported in a non-apoptotic type of programmed cell death seen during blow-fly salivary gland meta-morphosis (Bowen, Morgan and Mullarkey, 1993; Bowen, Mullarkey and Morgan, 1996). In this instance, *de novo* synthesis of mRNA gives rise to at least ten new proteins in a prelude to cell destruction. The first new proteins having a molecular weight between 30 and 100 kDa appear by day 8 of the life cycle and a number persist until the advent of cell death on day 9. New proteins appear in a cascade of production during day 8 and *in vitro* translation of mRNA produced at this time shows a signifi-cant 53–55 kDa protein appearing before cell destruction (Fig. 4.4).

Significantly, no evidence of DNA degeneration or laddering was observed; on the contrary, in this system considerable DNA synthesis in the form of endo-duplication was seen in the polytene chromosomes.

4.3.2 Programmed cell death in *Drosophila*

Most progress has been made using the fruit fly *Drosophila*, where a fuller genetic and chromosomal map is known. Much of our understand-ing of the operation of genes specifying spatial information relevant to body plan stems from pioneering work on *Drosophila* (Lawrence, 1973). There is also good evidence for local patterns of cell death in embryos mutant for segmentation genes (Ingham, Howard and Ish-Horowicz, 1985). Cell death occurring during embryological development in *Drosophila* (Abrams *et al.*, 1993) may be a response to hormonal cues or cell to cell interactions (Wolff and Ready, 1991). A number of mutations have been identified that have a bearing on the pattern of cell death seen during the development of particular tissues. One such mutant is *Fushi-taraza* (Margassi and Lawrence, 1988), where cell death occurs in alter-nate segments during development. Interestingly, here cell death appears to overspill to near neighbours and is not only confined to cells expressing the mutant gene. Another well-documented example involves the Bar-eye mutant, where, characteristically, the mutant has a strip of cells missing in the compound eye following an abnormal pattern of cell death in the embryonic and formative imaginal disks (Fristrom, 1972). Many genes thus affect tissue-specific signalling events which can be environmentally modulated.

4.3.3 *reaper* gene

A gene called *reaper* (*rpr*) that controls an apoptotic programmed cell death has been identified in *Drosophila* (White *et al.*, 1994). A small part of the third chromosome (75C1,2) is essential for all the cell deaths that occur during normal development. Although deletion of *rpr* protects embryos from X-irradiation-induced apoptosis, high doses of X-rays induce some apoptosis in mutant embryos, suggesting that the basic underlying cell death programme is still present though not normally activated in mutant embryos. The *rpr* gene product appears to be a small

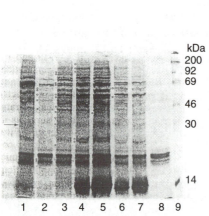

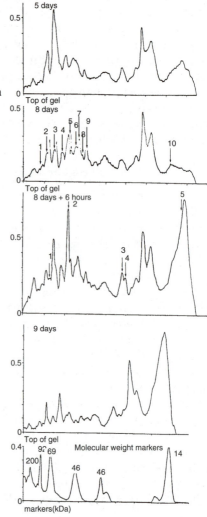

Proteins were synthesized from RNA extracted from: lane 1, the glands of 5-day-old larvae; lane 2, the glands of 7-day-old larvae; lane 3, the glands of 8-day-old larvae; lane 4, the glands of 8-day + 6 h old larvae; lane 5, the glands of 8-day + 12 h old larvae; lane 6, the glands of 8-day + 18 h old larvae; lane 7, the glands of 9-day-old larvae. Lane 8, no RNA; lane 9, molecular weight markers.

Fig. 4.4 Autoradiograph of an electrophoretic gel, showing the emergence of new proteins into metamorphic salivary gland cell death in the blow fly. A *de novo* isoenzyme of acid phosphatase also occurs (→). Densitometric scans of the gel indicate the appearance of a range of new proteins (↓1—10) into cell death between day 8 and 9 of larval life. One particular protein (↓2) at 53 kDa appears to peak immediately before the onset of cell death. Source: reproduced from Bowen, Morgan and Mullarkey, 1993, with permission.

peptide that shows little homology to known proteins, although it shows some similarity with mammalian death domains CD95 and TNFR1 (section 5.4.3). It is also not thought to be a cell death effector protein in that it is not part of the cell-death machinery itself. The gene and its product do clearly play a role in the activation of an apoptotic programmed cell death in that the cloned gene restores apoptosis to cell-death-defective embryos and *rpr* mRNA is specifically expressed in cells that are committed to death, and, indeed, the expression of *rpr*, precedes the first morphological signs of apoptosis by 1–2 h.

Experimental data (White *et al.*, 1994) indicate that the cell-death gene is located within an 85 kb interval of the H99 deletion of genomic region 75Cl,2 of chromosome 3 and encodes a single polyadenylated transcript of approximately 1300 nucleotides. Comparisons of homologous DNA segments from related *Drosophila* species suggest that the sequence is highly conserved for protein-coding purposes. Experiments demonstrate that the *rpr* mRNA transcript is expressed early in dying cells and indeed can also be recovered from phagocytosed corpses. The early expression of the transcript makes *rpr* mRNA an excellent specific marker for impending programmed cell death in *Drosophila*.

The absence of cell death in mutant embryos would clearly result in the persistence of extra cells and this in fact is found to be the case. The overall size of the central nervous system is virtually doubled and a two- to threefold increase in cell number is seen in the ventral nerve cord of mutant embryos that fail to hatch.

The identification of *rpr* marks just the beginning of a potentially explosive study of programmed cell death in *Drosophila* or, indeed, insect development generally. At this stage it is difficult to map exactly where the *rpr* gene might fit into the grand developmental cell death plan. Large numbers of cells undergo apoptosis during *Drosophila* and insect development and indeed specific programmed cell deaths also occur during metamorphosis. These deaths occur in a variety of locations and tissues at different times. The *rpr* gene, however, is the first gene discovered that appears to globally control cell death in *Drosophila*. In this regard, it is functionally comparable to *ced-3* and *ced-4* in the nematode *C. elegans* and yet it is not homologous. This multiple genetic control of entry into cell death does again imply that even more conserved genes, perhaps of a homeotic nature, must exist upstream. It could be argued, of course, that multiple pathways to programmed cell death would be desirable if not essential to achieve the necessary plasticity of response given the vagaries of environmental influences.

4.3.4 The *hid* gene

Another gene discovered in the genomic 75Cl,2 region is called *hid* (head involution defective) but appears to be a different kind of cell death regulator. *hid* function is required for some, but not all cell death expression in the embryo and developing eye of *Drosophila* (Hay, Wasserman and Rubin, 1995). In contrast with *rpr*, however, *hid* is expressed very broadly in non-dying cells as well, suggesting that *hid*-dependent cell death is regulated post-transcriptionally. Apoptosis induced by both genes is inhibited by baculoviral protein p35.

Hay, Wasserman and Rubin (1995) also showed that expression of the cell-death-regulatory protein REAPER (*RPR*) in *Drosophila* results in a small eye due to excessive cell death. It was shown that mutations in *thread* (*th*) enhanced *RPR*-induced cell death and that the gene *th* encoded a protein homologous to baculovirus inhibition of apoptosis proteins (IAPs).

4.4 BACULOVIRUS GENE PRODUCT INHIBITS APOPTOSIS IN INSECTS AND NEMATODES

The role of viruses and oncogenes in the inhibition of apoptosis is receiving urgent attention. It is clear that infected cells produce signals that flag their infected state to neighbouring cells and to the immune system. Thus, many viruses seem to have evolved mechanisms to interrupt or inhibit this signal, which would prove essential for the deletion of the infected cell. The virus has an interest in the continued life of the cell, not least as a source of viral replication and dispersal, and appears able to prevent the therapeutic suicide of the infected cell. A good example of this is seen in the inhibition of apoptosis in insect cells infected with baculovirus *Autorapha californica* (Clem, Fechheimer and Miller, 1991). Here, a specific gene product, p35, was identified as being responsible for blocking apoptosis. It has also been shown that expression of p35 in *C. elegans* prevents apoptosis and can rescue cells from mutant *ced-9*-induced cell death, suggesting that viral infection can inhibit normal apoptosis in different organisms and that p35 probably acts independently or downstream of *ced-9* in this pathway (Sugimoto, Freisen and Rothman, 1994).

Hay, Wasserman and Rubin (1995) called the *th* gene products that were homologous to the baculovirus inhibitors of apoptosis DIAP1 and DIAP2 (a related protein). Both suppressed naturally occurring apoptosis in *Drosophila* eye and inhibited both *rpr* and *hid* activity. They concluded that the IAP death-preventing activity was localized to the N-terminal baculovirus repeats, a motif that occurs in both viral and cellular apoptotic inhibitors (section 6.9.4).

4.4.1 Homology between insect and mammalian cell death domains

Emphasizing further the evolutionary conservation of the apoptotic machinery, the death protein RPR shares some homology with mammalian death domains in CD95 and TNFR1 and RPR mediates its action by activating ICE/ced-3-like proteases. It is thought that RPR may exert its actions by engaging ICE/Ced-3-like proteases in a similar manner to FADD and CD95 in mammals.

In addition it has also been shown that the 75 kDa tumour necrosis factor receptor (TNFR2) transduces extracellular signals via receptor-associated cytoplasmic proteins TRAF1 and TRAF2 (Chapter 5) and that TRAF2 has associated with it two proteins designated c-IAP1 and c-IAP2, closely related to mammalian members of the inhibitor of apoptosis protein (IAP) family, the family originally identified with baculoviral inhibition of apoptosis in insects.

B. ONCOGENES, TUMOUR SUPPRESSOR GENES AND PROTEASES THAT CONTROL APOPTOSIS

A range of oncogenes have now been identified that have a bearing on the control of apoptosis in mammals. **Oncogenes** are genes whose excessive activity provides the cell with selective growth advantage over other cells and thus could contribute to the formation of a tumour. **Tumour suppressor genes**, as the name implies, code for proteins that can suppress unrestrained cell growth through their specific inhibiting effects on the cell cycle and are involved in promoting apoptosis. The ICE/Ced-3 family of proteases has recently been shown to link cell death transduction mechanisms to cell nuclear collapse, DNA fragmentation and death.

4.5 BCL-2/BAX FAMILIES AND APOPTOSIS

The gene *bcl-2* (B-cell lymphoma 2) was first identified on human chromosomes 18 as the site of reciprocal translocation in follicular B-cell lymphoma. It encodes a membrane-associated protein, Bcl-2, present in the endoplasmic reticulum, nuclear and outer mitochondrial membranes. The anti-apoptotic activity of *bcl-2* was first noted when its expression was shown to prolong the survival of an interleukin-3-dependent myeloid (a collective term for non-lymphocyte blood cells) cell line after removal of the cytokine without inducing proliferation. In fact, *bcl-2* proto-oncogene protects cells from apoptosis induced by survival factor removal in many cell types, and is widely expressed during embryonic development. In adults it is restricted to immature and stem cell populations, in epithelia such as skin and intestine, long-lived cells such as memory-restoring B-lymphocytes, peripheral sensory neurones and glandular epithelial tissue.

bcl-2 synergizes with another oncogene, c-*myc*, in mammalian tumour progression, Bcl-2 suppressing c-*myc*-induced apoptosis (section 4.6) while leaving its proliferating activities unaffected. Bcl-2 expression also protects cells from the cytocidal effects of a variety of toxic agents. It does not appear to exclude cytotoxic drugs, nor does it confer resistance to DNA damage by genotoxic agents. It appears instead to act by suppressing the tendency of damaged cells to commit suicide. Thus, tumour cells that inappropriately express *bcl-2* may survive higher doses of chemotherapeutic agents *in vivo* despite sustaining significant genetic damage. In fact, it has been shown that anticancer drugs can induce cell-cycle arrest when the gene product Bcl-2 is present at high levels, but the cells typically fail to die or do so very slowly. Bcl-2 can convert anti-cancer drugs from cytotoxic to cytostatic, and cells containing Bcl-2 also survive gamma-irradiation. Thus cells with high levels of Bcl-2 will survive cancer treatments and resume their proliferation when drugs and/or radiation are withdrawn, producing highly resistant malignant clones (Chapter 6). Studies on many human tumours including neurob-lastoma, glioma, lymphoma, breast carcinoma, colorectal adenocarci-noma, prostate adenocarcinoma, melanoma and gastrointestinal malignancies have demonstrated a general correlation between increased expression of Bcl-2 (or Bcl-X$_L$) or decreased expression of the related protein Bax (a cell death promoter) and uncontrolled tumour cell growth, and, in certain cases, with tumour progression and poor progno-sis in cancer patients. For a review of Bcl-2 in cancer, see Reed (1996).

4.5.1 The Bcl-2 family of proteins and associated proteins

Bcl-2 is only one member of a still expanding family of proteins. The *C. elegans* protein Ced-9 is homologous with Bcl-2 and *bcl-2* is able to protect *ced-9* loss-of-function mutants from cell death. Thus, the struc-tures and function of *ced-9* and *bcl-2* have been conserved throughout metazoan evolution. The members of the family known to date are:

- **Bcl-2**: B-cell lymphoma 2
- **Ced-9**: cell death
- **Bcl-X$_S$** and **Bcl-X$_L$**: Bcl-2 homologue splice variants derived from the same gene
- **Bax**: Bcl-2-associated X protein
- **Bad**: Bcl-X$_L$/Bcl-2-associated death promoter homologue
- **Bak**: Bcl-2 homologous antagonist killer
- *A$_1$*: a novel haemopoietic specific early response gene
- *Mcl-1*: myeloblastic leukemia cell line gene
- **Viral proteins**:
  — p35: baculovirus
  — BHRF1: Epstein–Barr virus
  — VG16: Herpes saimiri
  — LMW5 HL: African swine fever virus
  — p19^{E1B}: adenovirus.

The data for this list are taken from Pearson *et al.* (1987); Clem, Fechheimer and Miller (1991); Neilan *et al.* (1993); Kozopas *et al.* (1993); Biose *et al.* (1993) and Lin *et al.* (1993).

Functionally, members of the family fall into two groups (Table 4.1). One group inhibits programmed cell death induced by growth factor deprivation, deregulation of *c-myc* or genotoxic damage, while the other group promotes cell death.

Furthermore, it is clear that the outcome in terms of cell death is dependent on how the members of this complex protein family combine and mix. The proteins can pair (**dimerize**). If the pairs are identical, this is called **homodimerization**. If the pairs are dissimilar, this is called **heterodimerization**. The outcome in terms of cell death will depend on whether the protein elements act as inhibitors or promoters of apoptosis, and on the way they mix.

Mammalian homologues of Bcl-2 have four conserved domains termed Bcl-2 domains (BD) — BH1, BH2, BH3 and BH4. Most members of the family contain a stretch of hydrophobic aminoacids at their C-terminus, which should allow for post-translational insertion into membranes. The Bad and A1 proteins, however, lack any obvious trans-membrane domains. Bax and Bcl-2 heterodimerize through their BH1 and BH2 domains to prevent cell death (Yin, Oltvai and Korsemeyer, 1994). Many of these protein interactions are seen in this family, some requiring both or only one of these domains, others, such as Bak and Bax, mediating protein-binding functions and apoptotic cell death through death effector domain BH3, a conserved domain distinct from BH1 and BH2. All anti-apoptotic members of the Bcl-2 protein family contain the BH4 domain (which is typically located near the N-terminus of these proteins; Zha *et al.*, 1996a). In contrast, the pro-apoptotic members of the Bcl-2 family lack BH4, with the exception of Bcl-X_S. Deletion mutants of Bcl-2 lacking the BH4 domain exhibit either loss of function or dominant inhibitory activity, paradoxically promoting apoptosis, thus indicating the functional significance of the BH4 domain (Hunter, Bond and

Table 4.1 Functional division of the Bcl-2 family of proteins

Cell-death inhibitors (anti-apoptotic)	Cell-death promoters (pro-apoptotic)
Bcl-2	Bax
Bcl-X_L	Bcl-X_S
Ced-9	Bak
Mcl-1	Bad
A1	
WR-13	
LMW5	
BHFR1	
p19^{E1B}	

Parslow, 1996). BH4 is not required for binding to Bax: the BH1 and BH2 domains carry this function (Hanada *et al.*, 1995).

Some of the family can form homodimers (e.g. Bcl-2, Bax, Bcl-X$_L$ and Bcl-X$_S$) and others, such as Bcl-2, can heterodimerize with Bax, Bcl-X$_S$, A1 and Bad. Bax can heterodimerize with Bcl-2, Bcl-X$_L$, Mcl-1 and A1. Bcl-X$_L$ can heterodimerize with Bax, Bad and Bcl-X$_S$ (Sato *et al.*, 1994; Farrow *et al.*, 1995; Kiefer *et al.*, 1995). Some cells preferentially use one member of their predominant survival factors; for example, peripheral blood lymphocytes primarily express Mcl-1. Eukaryotic cells clearly demonstrate tremendous redundancy in the expression of the Bcl-2 family members and the balance between anti-apoptotic and pro-apoptotic members varies as a result of homo- and heterodimerization reactions. For example, Bax appears to function as a homodimer to increase the sensitivity of cells to apoptotic stimuli. Binding of Bax by Bcl-2, Bcl-X$_L$ or Mcl-1 disrupts Bax/Bax homodimerization and protects cells from apoptosis.

Bad can form heterodimers with Bcl-2 and Bcl-X$_L$, mopping up these molecules and preventing them from forming heterodimers with Bax, thus allowing Bax to increase the sensitivity to cell death stimuli

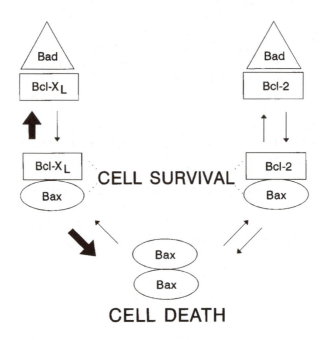

Fig. 4.5 Consequences of various Bcl-2 family homo- and heterodimerization reactions. Bax/Bax dimers promote apoptosis while Bcl-X$_L$/Bax dimers and Bcl-2/Bax dimers protect cells from apoptosis. Bad appears to counter the function of Bcl-X$_L$ and Bcl-2 by displacing them from Bax; this leads to an increase in Bax homodimers, promoting cell death.

(Fig. 4.5). These molecular controls of cell survival or cell death can be regarded as an autonomous rheostat controlling programmed cell death (Fig. 4.6).

Other proteins that bind Bcl-2 but share no significant homology with the Bcl-2 family of proteins include Bag-1, r-Ras, Raf-1, Nip1, Nip2 and Nip3. Nip1, Nip2 and Nip3 also bind to the adenovirus E1B 19 kDa protein (p19^{E1B}); the functions of these proteins are not clear (Boyd *et al.*, 1994). Raf-1 is best known for its role in the growth factor receptor-mediated signal transduction pathway, involving Ras, which is located in the plasma membrane (for further details on Ras-1 and Raf see Chapters 3 and 5). Bag-1 was shown to have anti-apoptotic activity, hence its name (Bcl-2-associated athano gene 1). In addition to binding to Bcl-2, Bag-1 can bind and activate Raf-1 kinase (Wang *et al.*, 1996). Bag-1 and Bcl-2 provide strong protection from apoptosis induced by anti-Fas antibody (see box) and cytolytic T cells (CTL) in certain cell lines. Bcl-2 alone only partially blocks apoptosis and Bag-1 by itself gives very little protection. Thus Bag-1 can functionally cooperate with Bcl-2, resulting in markedly more efficient suppression of apoptosis induced by anti-Fas antibodies and CTLs than either Bcl-2 or Bag-1 alone. These findings demonstrate that CTL and Fas cell killing is through a Bcl-2-dependent pathway, but adequate levels of an additional partner protein Bag-1 are required (Takayama *et al.*, 1995; see 4.5.2 for the possible mechanism underlying this cooperation). CTL killing is dealt with in section 5.7. For details of Fas-induced cell death, see box and section 5.4.3.

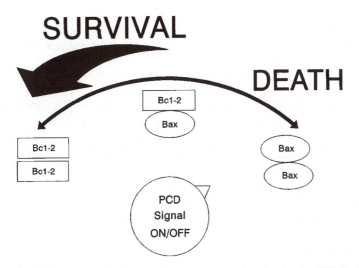

Fig. 4.6 Cell-autonomous rheostat of programmed cell death (PCD). Bcl-2 homodimerization or heterodimerization with Bax protects cells from apoptosis. Bax homodimers promote apoptosis.

The Fas-ligand (a protein) is located on the surface of cells and is a member of the TNF family (section 5.4.3). It binds to Fas receptors and in doing so can induce apoptosis. It has major roles in the immune system in downregulating immune reactions and in T-cell cytotoxicity.

The Fas receptor consists of 325 amino acids with a single receptor sequence at the N-terminal and membrane-spanning region, suggesting that Fas is a type 1 membrane protein. An antibody to Fas has been established called anti-Fas, which is an immunoglobulin M; in transformed mice cells it induces apoptosis within 5 h. Fas therefore transduces the apoptotic signal and the anti-Fas acts as an initiator. Cells that contain an intrinsic death programme are activated by Fas. There are similarities between Fas- and TNFa-mediated cell death. Loss-of-function mutations indicate that the *fas* gene may be a tumour suppressor gene, which suggests that the system plays a role in the onset of cancer. Cancer therapy could thus take the form of attempting to maintain the Fas system in transformed cells.

Several cytokines can modulate the levels of Bcl-2 and Bcl-X$_L$ in cells; these include IL-2, IL-3, IL-4, IL-6, IL-10, TGF-β and TNFα (section 5.8). A possible mechanism for IL-3 modulation of Bcl-X$_L$ is presented in section 4.5.2 below.

4.5.2 Possible mechanisms involved in controlling the Bcl-2 protein family

Although Bcl-2 and Bcl-X$_L$ are regarded as survival factors, the level of these proteins present in cells does not always indicate that the cell will survive when exposed to death stimuli. Recent publications have indicated that post-translational modification of Bcl-2-related proteins such as Bad may play a role in regulating their ability to promote cell survival.

It is the non-phosphorylated form of Bad that binds to membrane-bound Bcl-X$_L$, displacing Bax and promoting cell death (Yang *et al.*, 1995). Survival factors such as interleukin-3 (IL-3; section 5.8.2) trigger the phosphorylation of Bad, releasing it from BclX$_L$, which is membrane-bound, and allowing Bad to complex with a protein 14-3-3 that interacts with several signalling enzymes, including Raf-1 (Chapter 5). The phosphorylated Bad–14-3-3 complex resides in the cytosol; thus Bcl-X$_L$ is freed, enabling it to heterodimerize with Bax and promote survival (Zha *et al.*, 1996b; Fig. 4.7).

Withdrawal of IL-3 leads to Bad dephosphorylation and its release from 14-3-3 protein, again allowing it to bind to Bcl-X$_L$ and displace Bax. Bad appears to be a critical molecule in controlling cell death/cell survival by changes in its phosphorylation.

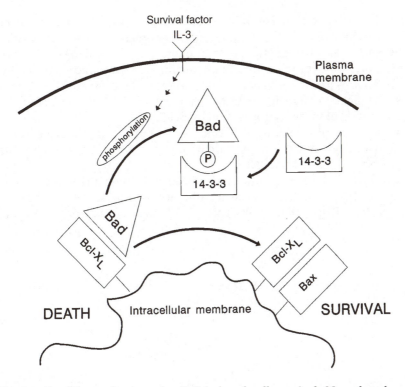

Fig. 4.7 Possible mechanisms for IL-3-induced cell survival. Non-phosphoryl-ated Bad heterodimerizes with membrane-bound Bcl-X$_L$. Phosphorylation of Bad leads to its release from Bcl-X$_L$ and allows it to complex with the protein 14-3-3 in the cytosol. This in turn allows Bcl-X$_L$ molecules to homodimerize or heterodimerize with Bax to promote cell survival. Source: redrawn from Zha *et al.* (1996b).

The ability of the chemotherapeutic agent taxol to induce phosphoryl-ation of Bcl-2 has been implicated as a mechanism by which it can promote death, even if there is high expression of Bcl-2 (Haldar, Jena and Croce, 1995). Consistent with this observation is the fact that deletion of the major serine/threonine phosphorylation sites of Bcl-2 enables it to promote cell survival under conditions where it is normally inactive (Cheng *et al.*, 1996). The kinase responsible for phosphorylation of Bcl-2 has not been elucidated. As previously mentioned, in addition to dimer-izing with other homologous proteins, Bcl-2 protein can associate with several non-homologous proteins, including the serine/threonine kinase Raf-1 and the GTPase r-Ras (Chapter 5). There is however, no evidence to indicate that Raf-1 is the kinase that phosphorylates Bcl-2.

Wang, Rapp and Reed (1996) have shown that Bcl-2 protein can target Raf-1 to mitochondrial membranes through an interaction that depends on the BH4 domain in Bcl-2 and the catalytic domain in Raf-1, promoting resistance to apoptosis. This allows Raf-1 to phosphorylate Bad or other proteins involved in the regulation of apoptosis. These similarities between Ras and Bcl-2 with regard to interactions with Raf-1 emphasize the importance of protein–protein interactions for the intracellular targeting of kinases to sites of biological importance. The competition between Ras and Bcl-2 protein for binding amounts of Raf-1 may explain the phenomenon of growth suppression by Bcl-2.

Wang *et al.* (1996) showed that Bag-1 can bind to and activate Raf-1; binding of Bag-1 to Bcl-2 is dependent on the BH4 domain, similar to the interaction with Raf-1 and Bcl-2.

Bag-1–Raf-1 complex may bind to the BH4 domain of Bcl-2, thus bringing Raf-1 to mitochondrial membranes associated with an activating protein Bag-1 (Fig. 4.8). This could explain the more efficient

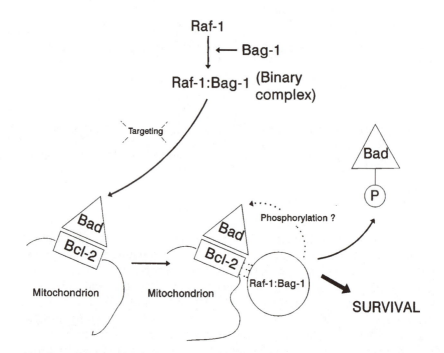

Fig. 4.8 Collaboration of Bcl-2 and Bag-1 (an anti-apoptotic protein) in the phosphorylation of Bad (a pro-apoptotic protein). Raf-1 and Bag-1 form a binary complex and are targeted to the mitochondrial membrane, where they bind with Bcl-2 and initiate the phosphorylation of Bad. Bad in its phosphorylated form is released from Bcl-2, allowing Bcl-2 to homodimerize or heterodimerize with Bax to promote cell survival. Bad may then complex with the protein 14-3-3. These events may be triggered by survival factors such as IL-3 (see Fig. 4.7). Source: redrawn from Gajewski and Thompson (1996).

suppression of cell death by anti-Fas antibody and CTL in the presence of Bag-1 and Bcl-2 compared with Bcl-2 alone (Chapter 4).

Since Bcl-2 can form dimers or even oligomers, Wang, Rapp and Reed (1996) suggest that it is possible that Raf-1–Bag-1 complexes assemble on these via BH4 domains on different Bcl-2 protein molecules. This would allow Raf-1 kinase to phosphorylate Bad, releasing Bad from Bcl-2 in a similar way to release of Bad from Bcl-X_L (Fig. 4.7), thus relieving repression of these anti-apoptotic proteins by allowing them to homodimerize with themselves or to interact with other proteins such as Bax (Fig. 4.8). Another possibility is that they would form membrane pores.

Minn *et al.* (1997) have demonstrated that Bcl-X_L forms ion channels in synthetic lipid membranes. The three-dimensional structure of Bcl-X_L is similar to the pore-forming domains of bacterial toxins, which form ion channels in biological membranes. Like the bacterial toxins, Bcl-X_L can insert into plasma lipid bilayers and form an ion-conducting channel. This channel is pH-sensitive and becomes cation-selective at physiological pH.

Bcl-X_L, like Bcl-2, has been localized in the outer mitochondrial membrane, outer nuclear envelope and the endoplasmic reticulum (ER). Bcl-2 can influence calcium distribution between the lumen of the ER and the cytoplasm. The outer membrane of the mitochondrion where Bcl-2 and Bcl-X_L are to be found is the site of numerous ion channels, and mitochondria store calcium ions if they are in excess in the cytoplasm. Nuclear pores collapse and nuclear transport is inhibited if calcium is removed from the intermembrane space of the nuclear envelope (Perez-Terzic, 1996). Thus, Minn *et al.* (1997) argue, all these sites have the potential to influence metabolic processes through the regulation of ion permeability. They also suggest that Bcl-X_L may also regulate protein passage in a similar way to diphtheria toxin. Liu *et al.* (1996) have suggested that cytochrome *c* (a protein situated in the mitochondrial intermembrane space), if released to the cytoplasm, could promote apoptosis through the activation of interleukin 1β converting enzyme (ICE)-like protease. Thus Minn *et al.* (1997) suggest a possible mechanism for Bcl-X_L maintenance of cell survival by blocking apoptosis through regulation of the permeability of the intracellular membranes where it resides.

Bcl-X_S is the only known pro-apoptotic member of the Bcl-2 family that contains a BH4 domain. Since Bcl-X_S can interfere with Raf-1, Wang, Rapp and Reed (1996) suggest that one mechanism by which Bcl-X_S may antagonize the function of Bcl-2 is by competing with Bcl-2 for binding to Raf-1. Although both Bcl-X_S and Bcl-2 reside in the outer mitochondrion membrane, only Bcl-2 can dimerize with Bad. The Bcl-2–Raf-1 interaction provides a new insight into the control of apoptosis, but it may not be essential for the function of Bcl-2 as a suppresser of cell death, although Raf-1 can at least modulate Bcl-2 function in some cells.

Bcl-2 and Bcl-X$_L$ appear to have two ways in which they can prevent cell death, through ion-channel formation and by interacting with other molecules. Recent studies have indicated that CED-4 and its mammalian equivalent Apaf-1 can biind to Bcl-2 and Bcl-X$_L$ (Chinnaiyan *et al.*, 1997b) as well as being involved in activating cell death proteases. This has led to speculation that CED-4/Apaf-1 mediate between anti-apoptotic molecules such as CED-9/Bcl-2 and cell-death promoting proteases such as CED-3/caspase-3 (see section 4.8 for details on the cell death proteases). Reed (1997) reviews the current information available with regard to the Bcl-2 family of proteins and their possible mode of action.

4.6 C-MYC AND APOPTOSIS

The c-*myc* proto-oncogene encodes a short-lived sequence-specific DNA-binding nuclear phosphoprotein (c-Myc), the expression of which is elevated or deregulated in virtually all tumours. Normally, c-*myc* expression is tightly regulated by mitogen availability. Although a peak of c-*myc* expression in fibroblasts is observed some 3 h after mitogenic stimulation, both c-*myc* mRNA and protein are continuously present at an appreciable level throughout the cell cycle in proliferating cells. As both c-*myc* mRNA and protein have very short half-lives in fibroblasts, this sustained presence of Myc protein can only result from continuous synthesis. The role of c-*myc* in cell growth is reviewed by Evan and Littlewood (1993).

The c-Myc protein appears to be a transcription factor and possesses an N-terminal domain with transcriptional activity and C-terminal DNA binding/dimerization basic helix–loop–helix leucine zipper (bHLH/LZ) domain, akin to that present in several known transcription factors. To date, however, the target genes have not been clearly defined. The c-Myc protein heterodimerizes with another bHLH/LZ protein partner, Max, and transactivates target gene expression upon binding to the DNA.

Deregulation of c-*myc* expression is associated with the inability to withdraw from the cell cycle and suppression of differentiation. Thus, c-*myc* probably encodes a transcription factor that targets growth-related genes promoting cell proliferation and suppressing arrest.

Given its growth-promoting and oncogenic properties, the observation that c-Myc also induces apoptosis was rather a surprise (Askew *et al.*, 1991; Evan *et al.*, 1992). High levels of expression of c-*myc* correlate both with increased proliferative rate and with increased sensitivity to apoptosis. Identical regions of the c-Myc protein are required for both growth promotion and apoptosis, and dimerization with the heterologous partner protein, Max, is necessary for both transforming and apoptotic functions of c-Myc. Amati *et al.* (1994) strongly suggest that c-

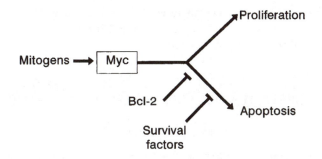

Fig. 4.9 Dual signal model for Myc-induced apoptosis. Mitogenic stimulation of Myc activates both growth promotion and apoptosis. Induction of apoptosis being an obligate function of Myc is regulated by the availability of anti-apoptotic survival factors (such as IL-3) and Bcl-2. Source: redrawn from Harrington, Fanidi and Evan (1994).

Myc induces apoptosis via a transcriptional mechanism, presumably by modulating appropriate target genes.

Various models have been proposed in an attempt to explain c-Myc-induced apoptosis in cells deprived of serum or treated with cytostatic agents. Support is growing for the 'dual signal model' (Evan and Littlewood, 1993; Harrington, Fanidi and Evan, 1994), which proposes that induction of apoptosis is a normal physiological function of c-Myc-induced proliferation. Hence, in order to grow, a cell requires two independent signals, one to trigger mitogenesis and the other to suppress apoptosis (Fig. 4.9).

It appears that c-Myc continually drives both a proliferative and an apoptotic programme. The reason c-Myc-expressing cells die more rapidly when treated with cytostatic and cytotoxic drugs is because they have a primed apoptotic pathway and are therefore poised to commit suicide in response to these substances. Any clone of cells that are proliferating is programmed to die if they outgrow the supply of survival factors. Most cancers, therefore, have had to evolve mechanisms to suppress apoptosis. A prime example is Bcl-2 activation, which mitigates the apoptotic effects of deregulated *c-myc* (Bissonnette *et al.*, 1992). Cells that have high levels of Bcl-2 or have lost p53 retain the ability to undergo apoptosis but have a raised threshold at which is its triggered. It has been proposed by Harrington, Fanidi and Evan (1994) that the basal machinery of apoptosis cannot be lost as it is entirely comprised of components essential to growth of metazoan cells. Thus, the cell's death and proliferative programmes are concomitantly activated so that every cell that enters the cell cycle can survive only as long as its death is suppressed by survival factors; this helps to prevent uncontrolled cell growth and formation of tumours.

For further information see Evan *et al.* (1994); Harrington, Fanidi and Evan (1994). These two review articles provide further details on the role of c-Myc in cell proliferation and cell death.

4.7 THE ROLE OF TUMOUR SUPPRESSOR GENES IN APOPTOSIS

4.7.1 p53 protein

The tumour suppressor phosphoprotein p53 has a molecular weight of 53 000 and is involved in mediating the cellular response to DNA damage and maintenance of genomic integrity. DNA–p53 interaction must occur for p53 to function as an activator or inhibitor of transcription or regulator of DNA synthesis and repair. The p53 protein is divided into three regions, the N-terminal transcription region, the central sequence-specific DNA-binding region and the C-terminal oligomerization region, which helps to stabilize the binding activity as well as recognizing DNA damage.

The p53 protein is normally short-lived, but when cells are exposed to DNA damaging agents (such as ultraviolet light, gamma-irradiation, genotoxic chemicals) it becomes stabilized and increases in concentration, or is activated. This leads to G_1 cell-cycle arrest, allowing the cell to repair damaged DNA before proceeding to the S-phase. If the DNA damage is irreparable, the increased activity levels of p53 direct competent cells to self-destruct by undergoing apoptosis.

The tumour suppressor gene *p53* mediates growth arrest through its role as a transcriptional activator; it induces the expression of a 21 kDa protein termed Waf-1 (wild-type p53-activated fragment) or Cip1 (CDK interacting protein), which interacts with and inhibits cyclin-kinase complexes (Chapter 3), thereby preventing cell-cycle progression; it also transactivates genes such as *GADD45* (a growth-arrest- and DNA-damage-responsive gene).

Waf-1/Cip1 transcriptional activation does not appear to be required in p53-induced apoptosis: in fact, p21 protein may be anti-apoptotic (Deng *et al.*, 1995; Brugarolas *et al.*, 1995). This is surprising as a number of inhibitors of kinases that progress the cell cycle appear to induce apoptosis; for example, inhibitors of $p34^{cdc2}$. Interestingly, perforin/granzyme-mediated cytotoxic T lymphocyte (CTL) killing requires target cells to be cycling, and cyclin-bound cdc2 and cdk2 kinases are activated during granzyme-induced apoptosis (section 5.7.2). There is obviously a need for further clarification with regard to inhibition or activation of kinases during apoptosis.

Recently, depletion of ribonucleotide (NTP) pools, even in the absence of detectable DNA damage, has been shown to increase p53 protein levels and initiate G_1 arrest (Linke *et al.*, 1996). Cytokine withdrawal also triggers the p53 apoptotic pathway (Fig. 4.10).

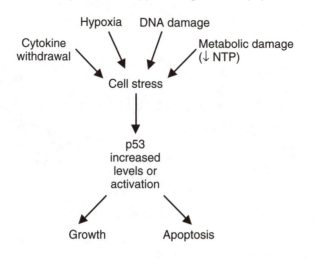

Fig. 4.10 Cytokine withdrawal, hypoxia, DNA damage, and depletion of ribonucleotide (NPT) pools increase p53 protein levels, leading to either growth arrest or apoptosis.

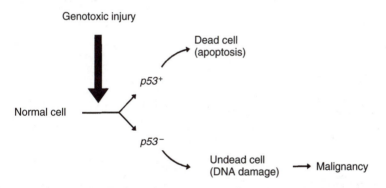

Fig. 4.11 Damaged cells that are irreparable undergo apoptosis if p53 is expressed by the cell. In p53-null cells, the DNA damage may lead to malignancy.

4.7.2 Tumour suppression by p53

Oncogenic mutations are reduced by p53 because of its ability to facilitate DNA repair. This indirect mechanism of tumour suppression is augmented by p53's ability to promote apoptosis. Loss of p53 function can thus contribute to tumorigenesis by allowing inappropriate cell survival (Figs 4.11, 4.12).

4.7.3 Tumorigenesis and p53

Apoptosis and necrosis both occur in tumours (Chapter 6). Mutations suppressing apoptosis could promote tumour progression by decreasing the cell loss factor.

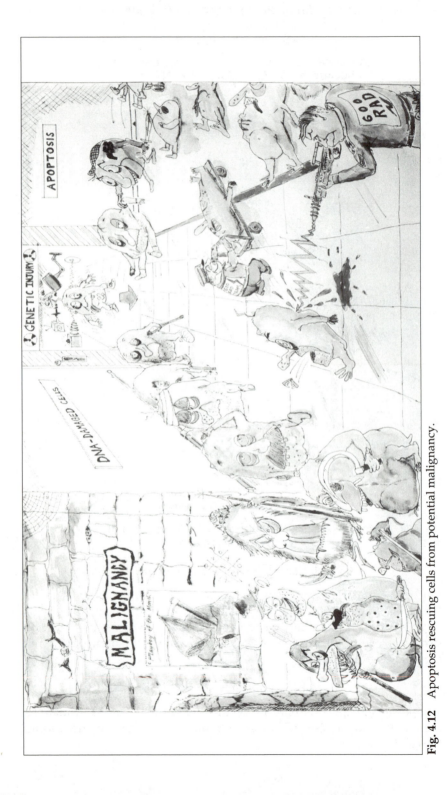

Fig. 4.12 Apoptosis rescuing cells from potential malignancy.

Table 4.2 The apoptotic index

	Cell loss (apoptotic index)
Colorectal tumours	0.96
Malignant melanoma	0.73

The apoptotic index (Table 4.2) can affect the kinetics of the tumour. The lower the index, the more aggressive the tumour.

The cytotoxicity of many anticancer chemotherapeutic agents may result from their ability to activate apoptosis. Consequently, mutations in the apoptosis programme (such as *p53* mutations) may reduce the effectiveness of cancer chemotherapy.

4.7.4 *p53* mutations and tumorigenesis

• *p53* deletions or mutations have been detected in over 50% of human cancers and occur in a wide range of tumour types.

• Li–Fraumeni syndrome (a familial cancer syndrome that predisposes individuals to a variety of tumours) has germ-line mutations in *p53*.

• *p53* mutations are often a late event in tumorigenesis.

• *p53* mutations are associated with clinically aggressive cancers and poor patient prognosis.

• Mutant mice homozygenes for *p53* deletions are predisposed to spontaneous and carcinogen-induced tumours. (**Note**: The mice develop normally; p53, therefore, has no essential role in physiological cell deaths that occur during embryogenesis. Nevertheless, endogenous p53 participates in apoptosis under physiological conditions.)

• *p53* mutations are rare in highly curable cancers, e.g. certain leukaemias and lymphomas, testicular cancer and Wilms' tumour. There is a Wilms' tumour subtype that responds poorly to chemotherapy; this subtype has *p53* mutations.

• Mutiple myeloma and acute lymphoblastic leukaemia often have *p53* mutations in relapse-phase tumours. Mutations in *p53* dramatically reduce the probability that patients with B-cell chronic lymphocytic leukaemia will undergo tumour regression and enter remission following chemotherapy. This shows a direct link between *p53* mutation and resistance to chemotherapy.

4.7.5 p53 antibody production

Antibodies to p53 are very rare in healthy individuals and the levels of p53 are also low. It has been shown that many cancer patients have p53 antibodies and most of these exhibit an accumulation of mutant p53 in their tumour cells (Soussi, 1996). In fact, up to 40% of such patients with an alteration in the *p53* gene develop p53 antibodies. This information is consistent with the fact that *p53* mutations are a predictor for unfavourable prognosis in various cancers, e.g. breast and colon cancers.

Interestingly, Yanuck *et al.* (1993) generated CTL response to cells containing mutated p53 isolated from lung cancer. These cells recognized not only the mutant p53 peptide but also target cells containing the peptide. It is not clear whether every type of *p53* mutation (there are more than 700 types) can lead to such a positive cellular response.

4.7.6 Mechanism of p53-mediated apoptosis

This mechanism is poorly understood and appears to vary depending on the cells or tissue examined; for instance, not all apoptotic pathways require p53 even within the same tissue. Thymus apoptosis can be divided into p53-dependent and p53-independent apoptosis. Gamma-radiation-induced apoptosis requires p53 while other apoptotic stimuli do not need p53, e.g. glucocorticoid-induced apoptosis in thymus (Clarke *et al.*, 1993). No new gene expression is apparently required for increased p53 levels following gamma-irradiation of thymocytes as actinomycin D does not prevent apoptosis, while glucocorticoid apoptosis is prevented by actinomycin D treatment. Serum depletion will eventually induce apoptosis in p53-deficient cells transformed by adenovirus protein E1A but much more slowly than if p53 was present. Chemotherapeutic agents also induce apoptosis in p53-deficient cells, although apoptosis typically requires considerably higher doses than in p53-expressing cells. The conclusion is that apoptosis can be executed without p53, but it is more difficult to trigger apoptosis in the absence of p53.

An interleukin-3-dependent murine lymphoid cell line (section 5.8.2) has been used to demonstrate that a survival factor (IL-3) can modify outcomes of p53 activation after cell irradiation. In the absence of IL-3, cells underwent apoptosis; in the presence of IL-3 they showed growth arrest (Collins *et al.*, 1992; Canman *et al.*, 1995). Interleukin-6 (section 5.8.3), which induces murine myeloid leukaemic M_1 cells to differentiate, can also prevent p53 induction of apoptosis in these cell lines (Yonish-Rouach *et al.*, 1991).

Apoptosis triggered by p53 can be inhibited by Bcl-2- but not p53-induced G_1 arrest, demonstrating that the two functions of p53 are independent. There are, apparently, transcriptionally dependent and independent mechanisms by which p53 controls apoptosis and this may be dependent on the cell type; or both mechanisms may occur in the same cell type (Yin *et al.*, 1997). Bax activation and/or Bcl-2 repression are examples of how p53 might shift the balance in favour of cell death. Bcl-2/Bax ratio is influenced by overexpression of *p53*; both a decrease in *bcl-2* and increase in *bax* message and protein levels are associated with p53-induced apoptosis (Fig. 4.5). Miyashita *et al.* (1994) showed that restoration of *p53* function resulted in rapid down regulation of Bcl-2 protein levels and apoptotic death. They also showed elevated levels of Bcl-2 protein in some tissues in *p53*-deficient transgenic mice as compared with the normal littermate control animals, which retained both copies of their *p53* gene. However, loss of *p53* did not detectably affect *bcl-2* expression in many tissues, implying that the extent to which basal levels of p53 influence *bcl-2* is highly tissue-specific. For example, *bcl-2* is not normally expressed in the liver and, in the absence of p53, it was still not expressed, implying the existence of a p53-independent mechanism for repression of *bcl-2*. In fact, a p53-independent negative regulatory element (NRE) has been described in the *bcl-2* gene. Thus, basal levels of p53 activity may be insufficient to significantly down-regulate *bcl-2* gene expression in some types of cells; however, the elevated levels of p53 activity associated with genotoxic stress could be important as an *in vivo* mechanism for downregulating *bcl-2* and inducing apoptosis. Gene transfer studies have demonstrated that enforced production of Bcl-2 protein at high levels can partially or completely block apoptosis induced by p53, suggesting a direct functional connection between p53 and its ability to both induce apoptosis and to downregulate *bcl-2* gene expression. DNA damage can induce apoptosis in proliferating lymphoid cells via p53-independent mechanisms inhibitible by Bcl-2 (Strasser *et al.*, 1994). Bcl-2 can also block both p53-dependent and p53-independent pathways of thymus apoptosis. Thus, Bcl-2 appears to function at a point distal to the convergence of p53-independent and p53-dependent limbs of a final common pathway for drug- and radiation-induced apoptotic cell death.

The tumour suppressor p53 is a direct transcriptional regulator of *bax* gene expression. It has been shown that p53 strongly transactivates the *bax* promoter and induces marked elevations in *bax* mRNA levels and Bax protein (Miyashita *et al.*, 1994). However, some cell systems have no upgrading of Bax during p53-induced apoptosis; furthermore, Bax-null mice undergo p53-dependent apoptosis (Knudson *et al.*, 1995). Much remains to be clarified with regard to the mechanism of p53-induced apoptosis. Recently, Yin *et al.* (1997) have shown that Bax is required for 50% of p53-induced apoptosis; the other 50% could result from p53 induction of transcription-independent mechanisms or regulation of other factors by p53.

The above studies suggest potential mechanisms by which p53 controls the sensitivity of cells to radiation-induced apoptosis, DNA-damaging chemotherapeutic drugs and even growth factor deprivation. The DNA damage produced by all these agents results in elevations in p53 protein and *p53* transcriptional activity. These elevations in p53 would be expected to induce an increase in Bax protein production and simultaneously a reduction in Bcl-2 protein synthesis. In doing so, p53 would shift the Bcl-2/Bax ratio into a condition of Bax excess and thus place the cell at increased risk of apoptosis.

4.7.7 Oncogene expression, p53 and apoptosis

Adenovirus E1A and E1B proteins are involved in oncogene transformation (Lowe, 1996). The *E1A* oncogene does not transform cells alone but collaborates with *E1B* or activated Ras to transform cells to a tumorigenic state. E1A promotes proliferation and entry into S phase and increases susceptibility to apoptosis. (**Note:** proliferation and apoptosis regions cannot be separated.) The tumour suppressor gene *p53* is required in E1A-promoted apoptosis. *E1B* encodes a 19 kDa protein (p19^{E1B}) and a 55 kDa protein (p55^{E1B}); these have no obvious effect on cell proliferation in the absence of E1A but they allow sustained proliferation of E1A-expressing cells. E1B counters the apoptotic effect of E1A through p55^{E1B} physically interacting with p53, converting its transactivation domain into a transrepressor, thus inhibiting apoptosis and promoting oncogenic transformation. p19^{E1B} (a functional homologue of Bcl-2) is also very effective at blocking E1A-induced apoptosis (Shen and Shenk, 1994). It is interesting to note that p19^{E1B} has been shown to inhibit apoptosis in response to TNFα or anti-Fas antibodies and Fas/Apo-1 (CD95) expression is induced by p53, linking it to the Fas ligand pathway of apoptosis (Owen-Schaub *et al.*, 1995). Chemotherapeutic agents and irradiation which activate p53 also enhance Fas/Apo-1 expression. Expression of the Fas/Apo-1 (CD95) TNF-receptor family transduces an apoptotic signal after ligand or specific antibody attachment (Nagata and Goldstein, 1995; Chapter 5). Thus *E1B* produces two proteins, p55^{E1B} and p19^{E1B}, that interact with different parts of the host cell apoptotic machinery. The protein p19^{E1B}, like Bcl-2, can block repression by p53 (Sabbatini *et al.*, 1995; Fig. 4.13).

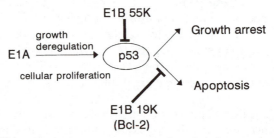

Fig. 4.13 *p53* is a tumour-suppressor gene that signals growth arrest or apoptosis. E1A and E1B are oncoproteins produced by adenovirus. E1A promotes growth deregulation and cellular proliferation, E1B 55 kDa protein can interact with p53 directly, preventing apoptosis, and E1B 19 kDa protein (homologous to Bcl-2) can block p53-induced apoptosis.

E1A-expressing cells contain metabolically stabilized p53 protein leading to a five- to tenfold increase in p53. This p53-dependent apoptosis is a cellular response to forced proliferation, which limits the oncogenic potential of E1A-expressing cells.

Other virus proteins that react like E1A include papillomavirus (HPV-E7) oncoprotein, and simian virus 40 (SV40) large T antigen. These viral oncoproteins are related to E1A in that they associate with a similar set of cellular proteins. The human papilloma virus HPV-E6 targets p53 for rapid destruction, promoting oncogenic transformation by HPV-E7.

4.7.8 Rb tumour-suppressor gene and its relationship with p53-induced apoptosis

The retinoblastoma gene (*Rb*) encodes a 105 kDa nuclear phosphoprotein termed pRb. In G_0 and early G_1 phases of the cell cycle, pRb is found in a hypophosphorylated form and becomes hyperphosphorylated throughout S, G_2 and most of M phase; as it emerges from M, phosphate groups are stripped (see Fig. 3.14). These phosphate groups play an important role in regulating pRb function (Hinds and Weinberg, 1994).

The DNA tumour virus oncoproteins (E1A + HPV-E7) succeed in eliminating pRb function by complexing with the hypophosphorylated forms of pRb, suggesting that this is the active form of pRb that blocks progression into late G_1 and S and hence cell proliferation (section 3.14). One consequence of Rb inactivation by viral oncoproteins is the constitutive activation of E2F transcription factor, which can then promote S-phase entry. The Rb protein is an inhibitor of cell-cycle progression at G_1/S by virtue of its ability to inhibit the activity of members of the E2F family of transcription factors (Chapter 3).

Loss of function of *Rb* allows cellular DNA synthesis during terminal differentiation of otherwise postmitotic lens fibre cells. This growth deregulation is counteracted by apoptosis switched on by p53 (White, 1994; Morgenbesser *et al.*, 1994). Hence, the loss of one tumour-suppressor gene is compensated by the activity of another, serving as a safeguard mechanism to control the emergence of neoplastic cell growth. Lens cells that would normally, in the presence of active Rb, withdraw from the cell cycle, elongate and differentiate into fibre cells, instead continue to synthesize DNA and die apoptotically. Thus, Rb implements the cell-cycle arrest required for normal differentiation, and continued progression of the cell cycle is incompatible with cell survival. DNA virus transforming proteins can, however, target both Rb and p53. Adenovirus E1A and HPV-E7 can interact with Rb at the protein level (E7 functionally inactivates p105^{Rb}), allowing E2F activation. Analogous

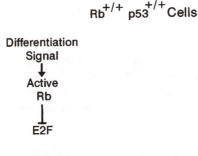

$Rb^{+/+}$ $p53^{+/+}$ Cells

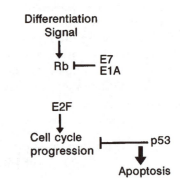

(a)

$Rb^{+/+}$ $p53^{-/-}$ Cells

Differentiation
Signal
↓
Rb⊢— E7
     E1A

E2F
↓
Cell cycle
progression
↓
Tumour

(b)

Fig. 4.14 (a) Response to differentiation signal in cells that express both Rb and p53. *Left*: a differentiation signal induces the formation of active Rb (hypophosphorylated form) which blocks the activation of E2F transcription factor, thus blocking cell-cycle progression. *Right*: Active Rb function can be eliminated by the virus proteins E7 (papilloma virus) and E1A (adenovirus). Inactivation of Rb results in the constitutive activation of E2F transcription factor, leading to cell-cycle progression. This inappropriate progression of the cell cycle in the presence of a differentiation signal can be halted by p53 induction of apoptosis. (b) Response to differentiation signal in cells that express Rb but do not express p53. In the absence of p53, if Rb function is blocked by virus proteins, the activation of E2F, which results in inappropriate cycling of the cells, gives rise to tumour formation.

results have been obtained using SV40 large T-antigen mutant (T121), which perturbs p105^{Rb} function but does not interact with p53.

The papilloma virus HPV-E6 protein and adenovirus E1B 55 kDa protein (p55^{E1B}) can block apoptosis in the situation of Rb loss and p53 activation, through direct interference with p53, whereas Bcl-2 and the adenovirus E1B 19 kDa protein (p19^{E1B}) over-ride apoptosis indirectly.

Thus, without Rb and p53, E2F activation stimulates cell proliferation, which is unobstructed, leading to tumour formation. Rb, unlike p53, is critical for normal embryonic development. Mice embryos with no Rb die at around 15 days post-coitus (Fig. 4.14).

If T121 is expressed in p53^{null} mice, highly malignant tumours are formed. In p53$^{+/null}$ heterozygote mice, low-grade tumours occur at first and eventually nodules of more malignant cells appear that are p53-deficient. This demonstrates that, in the absence of functional p105^{Rb}, loss of p53 is required for highly metastatic tumours to evolve. When p53 is functional, tumours may still arise but tumour growth is inhibited by apoptosis.

4.7.9 Mechanisms of inactivation of p53

• Homozyous deletions of p53.

• Inactivation by physical interaction with viral (e.g. p55^{E1B}) or cellular proteins (e.g. Mdm-2 cellular protein; Lin *et al.*, 1994). Mdm-2 is found in human sarcomas. These tumours might be controlled if a drug could be used to liberate Mdm-2 from p53.

• Factors acting downstream of p53 can inhibit apoptosis, including p19^{E1B} and Bcl-2.

In several instances, oncogenes that promote apoptosis collaborate with those that repress apoptosis in oncogenic transformation (e.g. adenovirus *E1A* and *E1B*, papilloma virus *E7* and *E6*, *c-myc* and *bcl-2*). Oncogenic transformation does not, however, require escape from apoptosis: it is the balance between cell division and apoptosis that is critical to tumour formation (Chapter 6).

4.7.10 Stages of p53-triggered apoptosis

1. **Priming**:
    (a) Genotoxic damage
    (b) Disruption of Rb function
    (c) Oncogenes (c-*myc*, E1A, *ras*)

2. **Triggering**: p53 increase/activation

3. **Execution**.

Oncogenes (*c-myc, E1A, ras*) all increase susceptibility to apoptosis and also promote proliferation. Interestingly, the regions promoting proliferation in these oncogenes (*c-myc, E1A, ras*) are identical to those promoting apoptosis. Priming for apoptosis appears to be linked to forced proliferation.

Note: Increased p53 is required to trigger apoptosis since normal cells express low levels but still grow and are viable. Functional activated p53 is not essential for apoptosis, but apoptosis is much less likely to occur in the absence of active p53. See Yonish-Rouach (1996) and White (1996) for recent reviews of p53 and apoptosis. New p53-dependent transcripts that may be death-specific have been found by Polyak *et al.* (1997).

4.8 PROTEASES AND THEIR ROLE IN APOPTOSIS

4.8.1 Introduction

The *C. elegans* gene *ced-3* (Ellis and Horovitz, 1986) is required for cells to undergo cell death (Hengartner and Horovitz, 1994a). The protein encoded by *ced-3* shows significant similarity to proteins that affect programmed cell death in vertebrates (see also *ced-9* and *bcl-2*, which code for survival factors; section 4.5). The homology of gene products involved in promoting and preventing cell death in nematodes and vertebrates indicates that the molecular cell death pathway has been conserved.

4.8.2 ICE protease (caspase-1)

The first of these Ced-3 homologous proteins to be discovered was the mammalian cysteine protease interleukin-1-β-converting enzyme (ICE; Yuan *et al.*, 1993), which recognizes and cleaves the 31 kDa IL-1β precursor molecule into the biologically active 17.5 kDa form. Mature ICE can also, through limited proteolysis, convert inactive proenzyme ICE to the active mature form (Thornberry *et al.*, 1992).

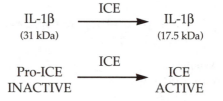

Overexpression of ICE in rat fibroblasts induces apoptosis (Miura *et al.*, 1993), suggesting that ICE functions as a cell death protein; yet

ICE-deficient mice undergo normal apoptosis, indicating that ICE might not after all be **the** mammalian cell-death protein. Elevated levels of other proteases can also cause apoptosis. The fact that ICE produces bioactive IL-1β and that cells secreting IL-1β do not automatically die (although often the producer cells undergo apoptosis) is puzzling. A possible explanation for a cell-death protein such as ICE activating IL-1β comes from Vaux (1996). He proposes that virus attacks on cells may activate ICE to convert the pro-IL-1β into bioactive IL-1β, which could diffuse from the cell, warning other cells with IL-1 receptors of a viral attack. IL-1β does act as an 'alarm' cytokine, activating lymphocytes and acting as a chemoattractant; it is also an endogenous pyrogen, increasing the temperature and blood flow during infection, causing inflammation. ICE could be used to mediate apoptosis in situations of viral attack, while other cell-death proteins homologous to Ced-3 could be the ones normally inducing apoptosis in development and tissue turnover. Other authors (Enari *et al.*, 1996; Fraser and Evan, 1996) regard ICE-protease (caspase-1) as a key player in apoptosis, inducing a cascade of proteases that have smaller proenzyme regions than ICE, such as Yama (caspase-3).

During the last few years, ten homologues of the *ced-3* gene product have been discovered. Unfortunately, different scientific workers have used different nomenclature, which has led to the same protein having several names (Table 4.3).

A general term now used for all the ICE-like family of aspartate-specific cysteine proteases (ASCP) is caspases (Alnemri *et al.*, 1996) — 'c' for cysteine protease mechanism and 'aspase' for their ability to cleave after aspartic acid. Each member of the family has been allocated an

Table 4.3 Nomenclature of the caspases (after Alnemri *et al.*, 1996)

ICE/CED-3 proteases	Caspase classification	Reference
ICE	Caspase-1	Yuan et al., 1993
ICH-1, Nedd-2	Caspase 2	Wang *et al.*, 1994; Kumar *et al.*, 1994
CPP32, Apopain, Yama	Caspase-3	Tewari *et al.*, 1995; Fernandes-Alnemri, Litwack and Alnemri, 1994; Nicholson *et al.*, 1995
ICH-2, TX, ICEre$_{II}$	Caspase-4	Kamens *et al.*, 1995; Munday *et al.*, 1995; Faucheu *et al.*, 1995
ICErel$_{III}$, TY	Caspase-5	Munday *et al.*, 1995
Mch2	Caspase-6	Fernandes-Alnemri, Litwack and Alnemri, 1995
ICE-LAP3, Mch3, CMH-1	Caspase-7	Duan *et al.*, 1996a; Fernandes-Alnemri *et al.*, 1995; Lippke *et al.*, 1996
FLICE, Mch5, MACH	Caspase-8	Muzio *et al.*, 1996
Mch6, ICE-LAP6	Caspase-9	Duan *et al.*, 1996b
Mch4	Caspase 10	Fernandes-Alnemri *et al.*, 1996

Arabic numeral based on its discovery and publication. All ten enzymes are synthesized as zymogens (proenzymes), which are proteolytically activated to form a heterodimeric catalytic region.

For details of each caspase, see the references in Table 4.3.

4.8.3 ICH-1/Nedd-2 (caspase-2)

Caspase-2 or ICH-1/Nedd-2 was the first Ced-3 homologous protein to be discovered after ICE. It is an embryonic brain protein and Wang *et al.* (1994) demonstrated that it was a cell-death protein. It is feasible that the large amount of neuronal cell death occurring during embryonic development is due to ICH-1/Nedd-2 (Chapters 3 and 6). Wang showed that *ICH-1/nedd-2* encodes two alternatively spliced mRNAs, *ICH-1$_L$* and *ICH-1$_s$*; the protein ICH-1$_L$ has 29% homology with Ced-3 and ICE and is 435 amino acids long. It is overexpression of ICH-1$_L$ that induces programmed cell death, while the truncated version ICH-1$_s$ (312 amino acids long) protects cells against death induced by serum deprivation. Unlike most other caspases, ICH-1 is not inhibited by the cowpox virus serpin cytokine response modifier A (CrmA), but Bcl-2 can inhibit ICH-1-induced apoptosis. (Serpins are serine protease inhibitors. Serine proteases have highly reactive serine residues in their active site.)

4.8.4 ICE protease inhibitors

CrmA inhibits most of the ICE family of aspartate-specific cysteine proteases (Ray *et al.*, 1992; Wang *et al.*, 1994) — caspase-2 being the exception — preventing apoptosis. It is also active against serine proteases such as granzyme B (Quan *et al.*, 1995), which is a protease that plays a key role in natural killer (NK) and cytotoxic T lymphocyte (CTL) cell killing. Perforin is involved in delivery of the granzyme to the target cell (Chapter 5). CrmA expression can in fact inhibit apoptosis in a wide range of systems, including that induced by activation of either Fas or the tumour necrosis factor receptor (TNFR; Enari, Hug and Nagata, 1995; Tewari and Dixit, 1995; Chapter 5).

Baculovirus p35 also inhibits caspases and has a broader spectrum of activity than CrmA, as it also inhibits caspase-2; it also blocks Fas- and TNF-induced cell death (Beidler *et al.*, 1995).

4.8.5 Yama/CPP-32/apopain (caspase-3), ICE-LAP3/Mch-3/CMH-1 (caspase-7) and Mch2 (caspase-6)

Caspase-3, caspase-7 and caspase-6 are the ICE-like homologues most closely related to Ced-3 and are synthesized as proenzymes, which are then processed to form active enzymes after cell death signalling; their pro-region is much smaller than that of caspase-1 and caspase-2. Caspase-3 and caspase-7 are made of two subunits derived from the precursor molecules procaspase-3 and procaspase-7. Fernandes-Alnemri *et al.* (1996) have shown that the caspase-3/p17 subunit and the caspase-7/p12 subunit can form an active heteromeric enzyme complex that can induce apoptosis. Caspase-3 can also cleave procaspase-7 into the active caspase-7, but not vice versa. The relationship between caspase-3 and caspase-7 is very close, with the activity of caspase-7 probably dependent on caspase-3 activity. Other ICE-like proteases may also depend on caspase-3 activity, with caspase-3 processing the proenzymes of caspase-6. Srinivasula *et al.* (1996) propose a protease cascade triggered by caspase-3: caspase-3 activation is a key event during apoptosis and it can also be initiated by granzyme B, an aspartate-specific serine protease (ASSP; Darmon, Nicholson and Bleackley, 1995; Martin *et al.*, 1996; Quan *et al.*, 1996), that is important in NK and CTL cell killing. (Aspartate-specific proteases cleave their protein substrates adjacent to aspartate molecules. The ICE cysteine proteases and granzyme B serine protease are unusual amongst eukaryotic proteases in that they prefer substrates with these acidic side-chains.)

4.8.6 Substrates of ICE-like proteases

Various apoptotic stimuli activate caspases, which, once activated, cleave the abundant 116 kDa nuclear protein poly-(ADP-ribose) polymerase (PARP) to yield an 85 kDa apoptotic fragment (see box).

POLY-(ADP-RIBOSE) POLYMERASE (PARP)

PARP is a highly conserved enzyme found in most eukaryotic nuclei in large quantities. In response to DNA damage such as ionizing radiation PARP binds to strand interruptions in DNA. The DNA-binding domain (DBD) of PARP recognizes DNA strand breaks through its zinc finger, leading to enzyme activation. Very large amounts of poly-(ADP-ribose) are rapidly synthesized in response to DNA strand breaks, leading to depletion of the NAD pool and transient addition of polymer to nuclear protein.

NAD+, and PARP and poly-(ADP-ribose) are covalently attached to various enzymes involved in DNA repair and replication and also to histones. Poly-ADP-ribosylization is probably the most drastic post-translational modification that occurs for both chromosomal proteins and nuclear enzymes involved in DNA metabolism.

DNA repair occurs after dissociation of PARP from DNA strand breaks. PARP itself does not participate directly in DNA repair. Lindahl *et al.* (1995) have reviewed PARP activity.

Evidence has been provided to indicate that nine caspases can directly cleave PARP: caspase-1, caspase-2 and caspase-4 (Gu *et al.*, 1995), caspase-3 (Tewari *et al.*, 1995), caspase-5 and caspase-6 (Ghayur *et al.*, 1997), caspase-7 (Singleton, Dixit and Feldman, 1996), caspase-8 (Muzio *et al.*, 1996) and caspase-9 (Duan *et al.*, 1996b).

Another substrate cleaved by caspase-3 is the catalytic subunit of the DNA-dependent protein kinase (DNA-PK; Casciola-Rosen *et al.*, 1996), cleavage occurring at a site that is highly similar to the cleavage site within PARP. Casciola-Rosen and colleagues propose that a central function of caspase-3, or similar caspases, is to cleave nuclear repair proteins. PARP, although a key target for caspases, appears to be just one of a variety of substrates attacked by this family of enzymes. Proteins involved in RNA splicing are also targets for protease attack during apoptosis (Casciola-Rosen *et al.*, 1996; Waterhouse *et al.*, 1996). Caspase-3 and caspase-7 and, to a lesser extent, caspase-6, but not caspase-2 and caspase-4 or CTL's protease granzyme B, were demonstrated by Waterhouse *et al.* (1996) to attack heterogenous nuclear ribonucleo-proteins.

Lazebnik *et al.* (1995) showed that caspase-6 cleaves nuclear lamins during apoptosis. If lamin proteinases are inhibited, apoptosis is blocked prior to packaging of the condensed chromatin into apoptotic bodies. Lazebnik and colleagues conclude that both independent pathways leading to nucleosomal fragmentation and lamin cleavage must work in parallel during the final stages of apoptosis and that neither pathway alone is sufficient to complete nuclear apoptosis. Instead, the various activities cooperate to drive the disassembly of the nucleus.

Yet another target for caspases such as caspase-3 could be fodrin. Fodrin is a cytoskeletal protein found in the cortex of many vertebrate cells; it is similar to the red cell membrane cytoskeletal protein spectrin. These proteins help to bundle and cross-link actin filaments in the cytoskeleton and play key roles in altering and stabilizing the shapes of many kinds of cells. It has been demonstrated that cleavage of alpha-fodrin accompanies apoptosis induced by Fas ligand (Martin *et al.*, 1995). Fas/Apo-1-mediated apoptosis has been shown to activate caspase-3 (Schlegel *et al.*, 1996), implicating this protease in Fas-induced cell death. Furthermore, CrmA, which inhibits caspase-3, substantially suppresses Fas-triggered cell death (Los *et al.*, 1995). Fodrin cleavage by caspase-3 or another ICE-type protease during apoptosis could implement the membrane blebbing seen during the cell-death process.

Different members of the family may preferentially attack different target proteins and all the target proteins may require simultaneous degradation for apoptosis to be successfully completed.

4.8.7 Non-activation of ICE-like proteases blocks apoptosis

Hyperosmotic stimulus in neurons leads to programmed cell death with a resulting fourfold decrease in Bcl-2 correlated with proteolytic processing of the caspases, caspase-3 and caspase-7. Singleton, Dixit and

Feldman (1996) have shown that type I insulin-like growth factor receptor (IGF-IR) activation blocks this osmotically triggered cell death by regulating apoptotic proteins. Interestingly, it prevents procaspase-3 and procaspase-7 processing, thus preventing the formation of the active proteases and also stabilizing Bcl-2. Increased IGF-IR expression enhanced the anti-apoptotic protein Bcl-X_L. Thus, IGF-IR activation can inhibit cell death by preventing activation of the protease mediators of cell death and by increasing the activity of anti-apoptotic proteins.

4.8.8 FADD-like/ICE-like proteases

Two recently discovered aspartate-specific cysteine proteases (ASCP) are Mch4 (caspase-10) and Mch5/FLICE/MACH (caspase-8; Fernandes-Alnemri *et al.*, 1996). They link the ICE/protease cell-death pathway directly to the Fas/Apo-1/CD95 death-inducing complex (section 5.4.3), which resides on the cell surface. The ICE family of proteases has been implicated in Fas-ligand and tumour necrosis factor (TNF)-induced apoptosis, since CrmA, which is a major inhibitor of the ICE cysteine proteases, was also shown to inhibit Fas- and TNF-induced cell death (Tewari and Dixit, 1995). Others have reported the requirement of active ICE proteases for Fas/Apo-1-mediated apoptosis (Los *et al.*, 1995). The transduction signal pathways from TNFα and Fas-ligand to their respective death receptors and death domains are explained in Chapter 5. Fas-associating protein with death domain (FADD) functions as the common signalling protein for cytokine-mediated cell death (Chinnaiyan *et al.*, 1996). Tumour necrosis factor receptor 1 (TNFR1)-associated death domain (TRADD) acts as an adapter molecule for TNFR1 (Hsu *et al.*, 1996), mediating the interaction of TNFR1 with FADD. The 117 N-terminal amino acids of FADD are capable of triggering apoptosis and consequently this region of FADD is called the death effector domain (DED; Chinnaiyan *et al.*, 1995).

Activation of the Fas/Apo-1/CD95 cell-death pathway initiates association with six proteins, called CAP by Kischkel *et al.* (1995); this stands for cytotoxicity-dependent Apo-1-associated proteins. CAP1 and CAP2 were identified by Kischkel *et al.* (1995) as serine-phosphorylated FADD. CAP3, 4, 5 and 6 appear to be part of an ICE-like cysteine protease. These six CAP proteins plus the receptor form a death-inducing signal complex (DISC; Peter *et al.*, 1996). This linking of cell-death membrane receptors to the cytosolic apoptotic cysteine proteases by the discovery of ICE-like proteases that have FADD-like domains has helped to explain the mechanism by which the TNFR/Fas pathway works. FLICE (FADD-like ICE), now called caspase-8 (Muzio *et al.*, 1996), and Mch4 (caspase-10; Fernandes-Alnemri *et al.*, 1996) are examples of these recently discovered proteases that have homology to both FADD and ICE-cysteine proteases. Caspase-8 contains two N-terminal stretches of approximately 60 amino

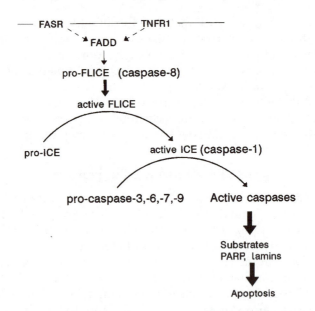

Fig. 4.15 The Fas receptor (FASR) and tumour necrosis factor receptor 1 (TNFR1) both trigger apoptosis via FADD (Fas-associating protein with death domain). FADD physically engages the apoptotic protease pro-FLICE (also called pro-caspase-8). The active protease FLICE can activate ICE (caspase-1) protease, which in turn may activate a cascade of apoptotic proteases including caspase-3, caspase-6, caspase-7 and caspase-9. Activation of caspases leads to the break-down of their substrates, which include PARP and lamins, resulting in the apop-totic death of the cell.

acids that are homologous to DED of FADD; the remainder of the protein is homologous to the active subunit of the ICE family (Muzio *et al.*, 1996).

Muzio and colleagues (1996) have clarified how the link between acti-vators of apoptosis and the effectors works. They demonstrated that

'a cell surface receptor (Fas) uses an adaptor molecule (FADD) to engage physically a cytosolic apoptotic protease termed FLICE. The death domain is an important signalling motif shared by both Fas and TNFR1 and oligomerization of this domain recruits cytosolic adaptor proteins to assemble a signal complex DISC. The assembly of a DISC is essential for Fas signal transduction. Upon activation, the death domain of the receptor binds to the death domain of the adaptor molecule FADD (CAP1-2) and thereby recruits it to form part of the DISC. The complete DISC is created when FADD, in turn, binds and recruits the ICE/Ced-3-like protease FLICE.'

Binding of FLICE (caspase-8) activates it, triggering PARP destruction and/or activation of other ICE proteases (Fig. 4.15).

For review of the ICE proteases and the role of FLICE (caspase-8), see Baringa (1996) and Fraser and Evan (1996).

Thome *et al.* (1997) have described a new family of viral inhibitors of apoptosis that inhibit the activation of FLICE. These inhibitors appear to have an effect on recruitment of CAP proteins into DISC, allowing recruitment of CAP1 and CAP2 but preventing recruitment of CAP3, 4, 5 and 6, which are essential for FLICE activation. They have named these inhibitors viral FLICE inhibiting proteins (v-FLIPs). FLIP, by preventing the recruitment of pro-FLICE into DISC, prevents the activation of FLICE by FADD.

4.9 CONCLUSION

A jigsaw of complex molecular information has become available during the 1990s relating molecules that initiate apoptosis to ones that transduce the message to the protein executors inside the cell. Other molecules, such as oncogenes and tumour suppressor gene products, are able to modulate the cell's response, depending on various external stimuli and which combination of molecules is present at any given time. An attempt has been made to give the reader a comprehensive overview of the information currently available on some of the key molecules involved in cell death and survival.

CHAPTER 5

Cytokines, signalling and cell death

This chapter will be divided into two sections. Section A will concentrate on the cytokines and their receptors as a method of signal transduction and signal reception respectively. Section B will consider how certain responsive cells such as natural killer cells and cytotoxic T lymphocytes effect an appropriate biological response. The role of various cytokines in protecting and initiating cell death will also be considered.

The philosophy of this chapter is reiterated in Fig. 5.1 for the convenience of the reader.

A. How cells respond to signals

5.1 OVERVIEW

In this section an attempt will be made to gain some understanding of the relationships between cytokine/growth factor receptor structure and

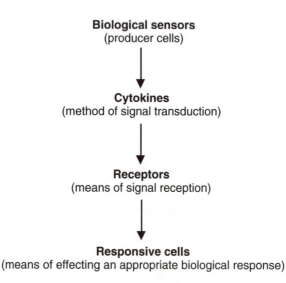

Biological sensors
(producer cells)

↓

Cytokines
(method of signal transduction)

↓

Receptors
(means of signal reception)

↓

Responsive cells
(means of effecting an appropriate biological response)

Fig. 5.1 The cytokine pathway.

function. Common themes are emerging in the intracellular pathways used by receptors of the haemopoietin/interferon family and by the receptor tyrosine kinases. Particular emphasis will be placed on the role of the MAP kinases in signal transduction, reflecting their importance in cell cycle and cell growth control. The identification and characterization of the direct upstream activators of MAP kinases has enlarged our understanding of phosphorylation networks.

The tyrosine kinase receptors for cytokines and growth factors activate MAP kinases by a complex mechanism involving the Src homology 2 and 3 (SH2- and SH3)-containing adaptor protein Grb-2, the GDP/GTP exchange protein m-Sos and a GTPase, Ras. The active GTP-bound Ras binds to Raf kinase and initiates the protein kinase cascade that leads to MAP kinase (ERK) activation. Activated MAP kinases translocate into the nucleus, where they interact with transcription factors, leading to the expression of certain genes.

Members of the Jun kinases (JNK)/stress-activated protein kinase (SAPK) family regulate pathways that can sensitize cells to apoptotic stimuli. These SAPKs or JNKs are a subgroup of the MAP kinase family.

5.2 INTRODUCTION

One of the ways in which higher eukaryotes receive messages from the extracellular environment is by means of receptors on the surface of cells.

The process of cellular interaction can be regarded as proceeding in two steps.

1. The extracellular molecule or ligand binds to a target cell because the latter possesses a specific receptor.
2. The receptor is activated as a result of ligand binding and intracellular biochemical pathways are stimulated leading to cellular responses such as cellular proliferation, differentiation and other effector functions.

The entire chain of events comprises signal transduction.

This chapter concentrates mainly on the cytokines as the functional ligands. These cytokines may be defined as secreted regulatory proteins that control growth, differentiation and survival of cells. They include families of regulators such as growth factors, colony-stimulating factors, interleukins, lymphokines, monokines and interferons. The monokines and lymphokines are involved in inflammation and immunity and colony-stimulating factors in haemopoiesis.

Cytokines seem to represent a somewhat bewildering collection of different molecules that often have overlapping biological activities in various cells and tissues and therefore a considerable degree of func-tional redundancy — many responses can be elicited by several different cytokines (Kishimoto, Taga and Akira, 1994). Most cytokines can also exhibit a wide range of biological effects, a phenomenon known as pleiotropy. Since many cytokine receptors have shared subunits (see later), the functional pleiotropy and redundancy is well explained.

By deleting the genes for certain of the cytokines it can be shown that few cytokines are absolutely essential for individual cellular function because one cytokine can compensate for the loss of another.

The cytokines are produced by cells that are widespread in the body. They therefore have many sites of production but few cellular targets and are rarely found in the circulation. The sphere of influence of the cytokines is limited to 'local' or autocrine/paracrine effects.

Cytokines play a key role in the development of all haemopoietic lineages, influencing cell survival, differentiation and cell death.

5.3 SH2/SH3- AND PH-SIGNALLING PROTEINS

Sequences referred to as Src homology 2 (SH2) and Src homology 3 (SH3) domains are small protein modules, 50 (for SH3 domains) to 100 (for SH2 domains) amino acid residues in length, that fold into compact domains. These domains mediate protein–protein interactions in signal transduction pathways that are activated as a result of ligand–receptor interactions. The SH2 domains bind to short, phosphotyrosine-containing sequences in, for example, growth factor receptors and other phosphoproteins, whereas SH3 domains bind to their target proteins through sequences containing prolyl and hydrophobic amino acid residues (Pawson, 1995). Proteins that contain SH2 and SH3 domains, e.g. Grb-2 (growth factor receptor-bound 2), use these modules in order to link the cytoplasmic domain of receptors to various cytoplasmic signalling pathways that eventually link the cell surface with the nucleus (sections 4.8.7, 5.4.2). Combinations of SH2 and SH3 domains are frequently found within the same polypeptide; for instance, Grb-2 is composed entirely of SH2 and SH3 domains (one SH2 and two SH3 domains) and functions as a molecular adapter to recruit other proteins to form signalling complexes and thus to nucleate a signalling cascade.

The plextrin homology (PH) domain was first identified as an internal repeat of about 100 amino acid residues in plextrin, the major substrate for protein kinase C in platelets. In addition to plextrin, the PH domain is found in a broad array of signalling proteins (at least 45), including all phospholipase C (PLC) isoenzymes, beta-adrenergic receptor kinase (βARK) and Bruton tyrosine kinase (Musacchio *et al.*, 1993).

A mutation in the PH domain of dynamin, a substrate for Bruton tyrosine kinase, is associated with X-linked agammaglobulinaemia, in which there is a profound decrease of mature B cells due to a block in B-cell development.

Most proteins that contain PH domains must associate with membranes in order to function; for instance, removal of the PH domain from βARK decreases kinase activity to basal levels (Fig. 5.2).

The PH domain is shaped like a β-barrel, a topology found in a class of proteins that includes such members as the bilin-, fatty acid- and retinol-binding proteins. The most notable feature of this class of proteins is that they all bind small lipophilic molecules in a cavity within the hydrophobic core of the barrel. The PH domains probably bind similar small molecules, such as the membrane phospholipid phosphatidylinositol-4,5-bisphosphate (PIP_2). The PH domains may therefore help to localize proteins to phospholipid membranes by interaction with phospholipid.

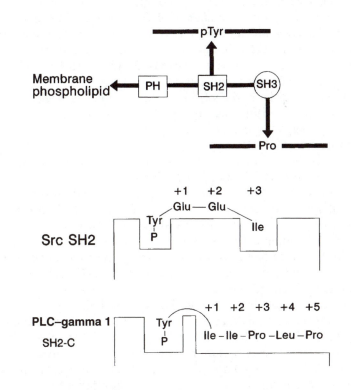

Fig. 5.2 (a) A protein containing SH2, SH3 and PH domains can potentially form multiple protein complexes contributing to networks of interacting proteins. (b) SH2 domains recognize pTyr-containing sites. Phosphorylation regulates binding, whereas the residues on the C-terminal side of the pTyr sites impart binding specificity (Pawson, 1995).

5.4 CYTOKINE RECEPTORS

The amino acid sequence (primary structure)-predicted higher orders of structure (secondary and tertiary structure) and biochemical functions suggest that there are at least four families of cytokine receptors:

• the haemopoietin/interferon receptors;
• receptors that have intrinsic tyrosine kinase activity (RTKs);
• the nerve growth factor/tumour necrosis factor (NGF/TNF) receptor family;
• G-protein-coupled receptors.

5.4.1 Haemopoietin/interferon receptors

The haemopoietin/interferon receptors are glycoproteins that span the cell membrane and have extracellular, transmembrane and cytoplasmic regions. The extracellular region is composed of a number of sub-domains, which may include a haemopoietin receptor SD100 domain, an interferon SD100 domain, a fibronectin type III domain, an immunoglob-ulin-like domain and an IL-2 receptor alpha chain domain. These domains may be represented in various numbers and combinations depending on the specific receptor type.

As mentioned earlier in the introduction, it is the presence of common receptor subunits or domains among the different cytokine receptor systems that helps to explain both pleiotropy and redundancy of cytokine function (Fig. 5.3).

The ligands that interact with the haemopoietin/interferon receptors include the interferons (IFNs) α, β and γ, interleukins (IL) 2, 3, 4, 5, 6, 7, 9, 10, 11, 13, leukemia inhibitory factor (LIF), oncostatin-M (OSM), erythro-poietin, ciliary neurotrophic factor (CNTF), growth hormone and prolactin (Gearing and Ziegler, 1993; Kishimoto, Taga and Akira, 1994).

Most of the haemopoietin/interferon receptors consist of two or more distinct subunits yet none possesses any intrinsic tyrosine kinase activity. When ligands bind to members of the haemopoietic/interferon receptor family, homo- or heterodimerization (and in some cases trimerization) of these receptors occurs, thus allowing their cytoplasmic regions to associ-ate with cytoplasmic signalling molecules such as members of the Janus kinase (Jak) family of protein tyrosine kinases, which are then activated, possibly by cross-phosphorylation. Oligomerization (dimerization or trimerization) of receptor components seems to be required for this asso-ciation and subsequent activation (cross-phosphorylation) to occur. It is possible that receptor components are constitutively associated with members of the Jak family and that the amount of associated Jak may increase after ligand binding (Fig. 5.4).

Depending upon the cytokine/receptor system, one or more of the

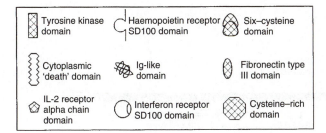

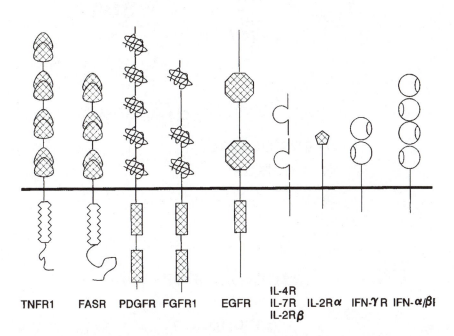

Fig. 5.3 Schematic view of receptor families showing shared receptor subunits that may contribute to cytokine redundancy and pleotropy. TNFR1 = tumour necrosis factor receptor 1; FASR = FAS receptor; PDGFR = platelet-derived growth factor receptor; FGFR = fibroblast growth factor receptor; EGFR = epidermal growth factor receptor; IL = interleukin; IFN = interferon.

four known Jaks (Jak1, Jak2, Jak3, Tyk2) is involved (Silvennoinen *et al.*, 1993). The activated Jaks phosphorylate both themselves and the receptor subunits on tyrosyl residues, thus creating docking sites for SH2-containing proteins that include SHC (Src homology/collagen), which couples receptor engagement to activation of the *ras* pathway.

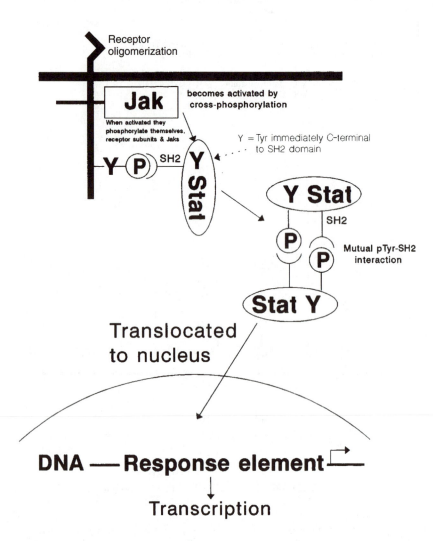

Fig. 5.4 Activation of SH2-signalling proteins by haemopoietin/interferon receptors. The activated complexes associate with Stat proteins, which become phosphorylated on tyrosyl residues. Phosphorylated Stat proteins form dimers by mutual pTyr–SH interaction and translocate to the nucleus, where they activate transcription by binding to DNA response elements.

The adapter molecule SHC plays a critical role in IL-2-induced activation of Ras. After IL-2 stimulation, SHC is tyrosine-phosphorylated and recruits Grb-2 to the tyrosine-phosphorylated IL-2 receptor component (IL-2Rb).

The activated Jaks can also phosphorylate one or more members of a family of signal transducers and activators of transcription (Stats). Each subset of the haemopoietin/interferon receptors regulates a unique set of Stat proteins. The phosphorylated Stats dimerize through mutual pTyr–SH2 interaction and are translocated to the nucleus, where they interact with DNA; for example, alpha-interferon (IFN-α)-induced transcriptional activation requires a complex of DNA-binding proteins, including tyrosine–phosphorylated Stat1 and Stat2 heterodimers, and p48, a protein that is not phosphorylated on tyrosine and comes from a separate family of DNA-binding proteins. In addition to these protein kinases, a lipid kinase, phosphatidylinositol 3-kinase (PI3-kinase) may also be activated, e.g. in the stimulation of T cells by IL-2.

The PI3-kinase consists of an 85 kDa regulatory subunit possessing SH2 domains that may facilitate recruitment and activation, and a 110 kDa catalytic subunit that possesses kinase activity. Addition of the potent and specific PI3-kinase inhibitor wortmannin to exponentially growing cells results in an accumulation of cells in the G_0/G_1 phase of the cell cycle. Wortmannin also partially suppresses IL-2-induced S-phase entry in G_1-synchronized cells; other SH2-containing proteins such as PLC-γ_1, Syp and PTP-1C may also be recruited by haemopoietin/interferon receptors.

5.4.2 Receptors that have intrinsic tyrosine kinase activity (RTKs)

More than 50 receptor tyrosine kinases are known and can be divided into at least 14 distinct subclasses on the basis of their extracellular ligand-binding domains. These domains may consist of immunoglobulin (Ig)-like motifs, fibronectin type III domains, cysteine-rich repeats, cysteine-rich regions and leucine-rich regions in various numbers and combinations; for example, the epidermal growth factor receptor (EGFR) extracellular region consists of two cysteine-rich repeats whereas platelet-derived growth factor receptors α and β (PDGFR-α and -β) consist of five Ig-like extracellular domains.

PDGF is a homo- or heterodimeric protein consisting of alpha and beta chains joined by disulphide bonds, giving rise to aa, bb or ab dimers. The PDGF-a receptor (PDGFR-a) binds both alpha and beta chains of PDGF, whereas the beta receptors bind only PDGF-b chains. Some cells, such as fibroblasts and smooth muscle cells, express both alpha and beta receptors whereas mesothelial cells express only alpha receptors. Neurons and capillary endothelial cells express only the beta receptors.

With one or two exceptions — for example, the insulin receptor, which exists constitutively as a heterotetrameric glycoprotein consisting of two extracellular alpha subunits and two membrane-spanning beta subunits joined by three disulphide bonds — the RTKs are single transmembrane glycoproteins with extracellular domains (see above), a transmembrane region and cytoplasmic domains. The cytoplasmic domains are catalytic in nature with latent tyrosine kinase activity. (Receptors for activin, inhibin and TGF-β have cytoplasmic domains that possess latent serine/threonine kinase activity instead of tyrosine kinase activity.)

Binding of the appropriate ligand to its receptor is accompanied by receptor homodimerization (Ullrich and Schlessinger, 1990), which leads to the juxtaposition of two cytoplasmic domains and activation of their intrinsic kinase activity, leading to phosphorylation of tyrosyl residues.

These tyrosyl phosphorylations probably proceed by a transphosphorylation mechanism whereby one receptor phosphorylates the other in the dimer. In the receptors for epidermal growth factor and platelet-derived growth factor (EGFR and PDGFR respectively), phosphorylation sites generally lie in non-catalytic regions of the cytoplasmic domain.

Multiple phosphorylation sites may exist in the cytoplasmic domains of RTK receptors each of which is relatively specific for a particular SH2-containing protein; for instance, phosphorylation of tyrosine 1021 in the receptor tail of PDGFR-β induces binding of PLC-γ_1 and thus stimulates the hydrolysis of phosphatidylinositol-4,5-bisphosphate (PIP_2) to inositol-1,4,5-triphosphate (IP_3) and diacylglycerol (DAG) (Fig. 5.5). Other phosphorylation sites are responsible for binding the p85 regulatory subunit of phosphatidylinositol-3-kinase (PI3 kinase; Tyr 740 and Tyr 751), members of the Src family of kinases (Tyr 579 and Tyr 581), SHC, Grb-2 (Tyr 716) and protein tyrosine phosphatase 1D (Tyr 1009; Fig. 5.6). (PI3 kinase catalyses the addition of phosphate to the 3' position of phosphatidyl inositol and other polyphosphoinositols.)

Grb-2 is constitutively bound to a second protein, the mammalian equivalent of son of sevenless (m-Sos), through its SH3 domain. Sos, in turn, interacts with Ras, which is bound to the cell membrane, and promotes the exchange of GTP for bound GDP (Egan *et al.*, 1993). Sos thus acts as a guanine nucleotide exchange factor.

Binding of GTP to Ras leads to the activation of the Ras molecule, which eventually results in the activation of MAP kinase by sequential protein kinase cascade reactions.

Activated MAP kinase translocates to the nucleus, where it participates in transcriptional regulation through the phosphorylation of transcription factors.

5.4.3 The nerve growth factor/tumour necrosis factor (NGF/TNF) receptor family

The defining features of the NGF/TNF receptor family are to be found in the extracellular domain, where multiple copies of domains containing six cysteine residues are to be found (Bazan, 1993).

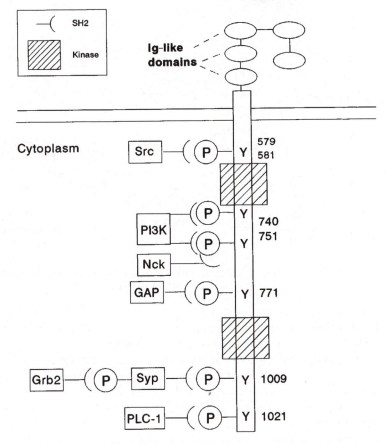

Fig. 5.5 Binding of molecules involved in signal transduction to autophosphory-lation sites of the PDGF receptor. Src = members of the Src family of tyrosine kinases; p85 and p100 = regulatory and catalytic subunits respectively of the phosphatidylinositol-3-kinase; GAP = GTPase activating protein of *ras*; PTP1D = phosphotyrosine phosphatase 1D; PLC-γ = phospholipase Cγ.

The cytokines TNFα and Fas$_L$ (the Fas ligand) exert their cell-killing effect by binding to the cell surface receptors TNFR1 and Fas respectively. The Fas receptor is also known as CD95/Apo-1 and plays an important role in the immune system as a mediator of T-cell death during the removal of autoreactive T cells and for the maintenance of immune system homeostasis. A CD40 receptor is expressed on B cells

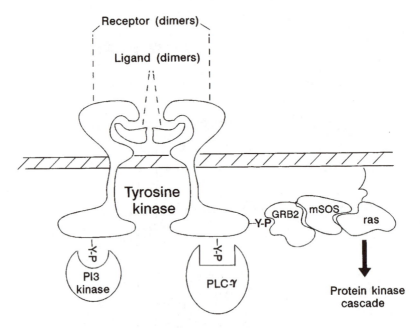

Fig. 5.6 A model for ligand-induced receptor dimerization and autophosphorylation and the link between receptor-dependent activation of Ras.

and is involved in immunoglobulin class-switching. (The TNF receptor 2 (TNFR2) is involved in thymocyte proliferation.)

The cytoplasmic regions of TNFR1, Fas and NGFRp75 contain a conserved apoptosis or 'death' domain of about 80 amino acid residues, which is an essential component of the signalling pathway that triggers apoptosis. The cytoplasmic regions of these receptors show no known enzymatic activity.

> The structure of the Fas 'death' domain has recently been eluci-
> dated by NMR analysis and consists of six antiparallel alpha-helices
> arranged in a novel fold. Helices 2 and 3 mediate the domain's abil-
> ity to self-associate, while binding to downstream effectors requires
> the region containing helices 5 and 6 (Huang *et al.*, 1996).

Both TNFR1 and TNFR2, when ligand-bound, activate the transcription factor NF-κB, leading to gene induction; however, the induction of apoptosis is largely mediated via TNFR1.

The transcription factor NF-κB exists as a cytoplasmic complex in most cells and in its quiescent state is bound to an inhibitory protein IκB.

When TNFR1 is activated as a result of ligand binding, IκB is phosphorylated and degraded and the released NF-κB migrates to the nucleus where it binds to κB elements on DNA, leading to transcriptional activation. NF-κB is also activated by many other pro-inflammatory and immunomodulator molecules, including IL-1, lipopolysaccharide, CD30 and CD40 ligands, lymphotoxin-α (LT-α) and LT-β. The lymphotoxins LT-α and LT-β belong to the TNF receptor superfamily (Zheng-gang *et al.*, 1996).

The NF-kB transcription factor is a key element of the rapid-response system that allows cells to quickly adjust gene transcription in response to new challenges. It was discovered as a protein nuclear factor (NF) that was able to bind to the immunoglobulin kappa light-chain transcriptional enhancer that is active in B lymphocytes (kB).

How are ligand/receptor binding signals transmitted to the activation of NF-κB and other downstream elements? The death domain seems to modulate receptor dimerization and multimeric interactions between receptor-associated proteins such as TRAFs (TNF receptor-associated factors), TRADD (TNF receptor-associated death domain), FADD/MORT1 (Fas-associated death domain) and RIP (receptor interaction protein; Fig. 5.7).

The Fas receptor recruits FADD/MORT1 while TNFR1 binds TRADD, which then acts as an adapter protein to recruit FADD/MORT. It therefore seems that FADD/MORT is a point of convergence between the Fas receptor and TNFR1 death pathways (Hsu *et al.*, 1996; Fig. 5.7). Details of the downstream caspase cascade from the Fas and TNF1 receptors, leading to apoptosis, have been fully addressed in Chapter 4.

TRAF2 is required for NF-κB and JNK activation signalled by the two TNF receptors (TNFR1 and TNFR2) as well as CD40. TNFR2 and CD40 do not, however, recruit FADD and so do not promote apoptosis (Fig. 5.8).

TRAF5 may be responsible for NF-κB activation signalled by the LT-β receptor (LT-βR). The TRAF family members are signal-transduction proteins.

TRADD interacts strongly with RIP, which is recruited to TNFR1 when the latter binds TNF. RIP contains an N-terminal region with homology to protein kinase C and a C-terminal region containing a 'death' domain. Transient overexpression of RIP in transfected cells leads to morphological changes characteristic of apoptosis, thus indicating that RIP is an apoptosis-inducing protein.

5.4.4 G-protein-coupled receptors

This group of cell membrane receptors, which typically have a seven-helix transmembrane (7TM) or 'serpentine' region, is by far the largest, consisting of hundreds of members. G-protein-coupled receptors are involved in responses to an extensive range of stimuli including light,

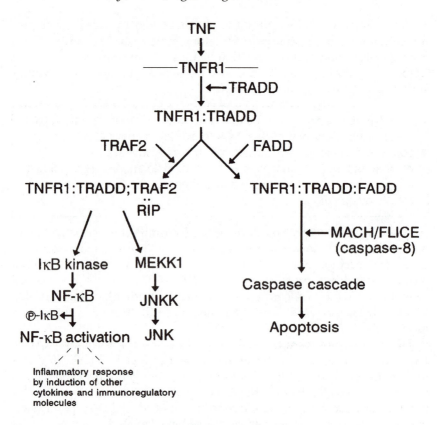

Fig. 5.7 Signal transduction from TNFR1. See text for details.

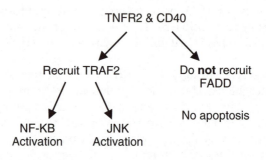

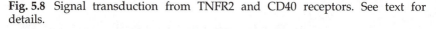

Fig. 5.8 Signal transduction from TNFR2 and CD40 receptors. See text for details.

smell, taste, hormones, pheromones and neurotransmitters (Bourne, Sanders and McCormick, 1991). The 7TM receptors have three extramembranous loops on either side of the plasma membrane and a cytoplasmic C-terminal tail (Fig. 5.9).

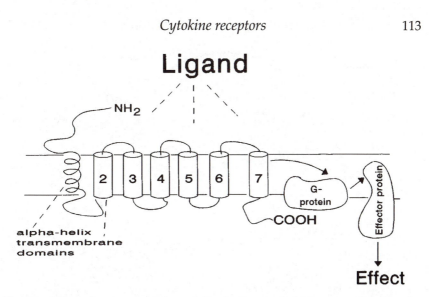

Fig. 5.9 Schematic representation of G-protein-coupled receptors showing the seven alpha-helical membrane-spanning domains. Binding of the ligand induces a change in the conformation of the receptor protein, enabling the latter to interact with a G-protein (guanyl nucleotide-binding protein). Association of ligand with receptor activates the G-protein, leading in turn to activation of effector proteins.

Upon ligand binding, the cytoplasmic portion of the activated receptor interacts with G-proteins. These G-proteins are heterotrimers, made from alpha, beta and gamma subunits (Gα, Gβ, Gγ), which serve to couple the seven transmembrane-type receptors to intracellular second messenger effector enzymes. The alpha subunit of the G proteins determines the specificity of the G-protein receptor interaction and is of fundamental importance for maintaining the fidelity of signal transduction. There are at least 20 vertebrate G-protein alpha-genes and several different beta and gamma subunits, giving rise to a large number of $\alpha\beta\gamma$ combinations. Each receptor only activates a specific subset of G-proteins and each G-protein only activates (or inhibits) a specific subset of effectors. Studies have suggested that a region of the G-protein adjacent to the C-terminal region plays a major role in effector interaction (Fig. 5.10).

Binding of ligands such as polypeptide hormones and neurotransmitters to 7TM receptors coupled to G-proteins activates adenylate cyclase and induces elevation of intracellular cAMP which results in the activation of protein kinase A (PKA), the catalytic component of which translocates to the nucleus. Nuclear PKA phosphorylates a cAMP response element binding protein (CREB), which is already bound to the cAMP response element (CRE). Phosphorylation of CREB increases its transactivation potential.

Cytokines, signalling and cell death

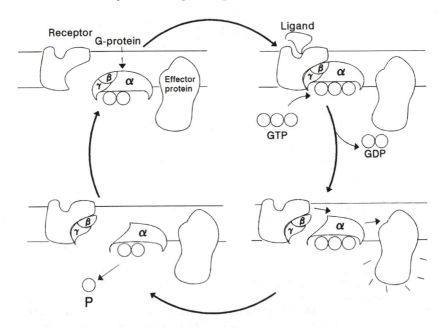

Fig. 5.10 G-protein-mediated effect of ligand binding. Ligand binding to the receptor leads to a change in receptor protein conformation. This change propagates to the G-protein; the alpha subunit exchanges GDP for GTP, then dissociates from the two other subunits, associates with an effector protein, and alters its functional state. The alpha subunit slowly hydrolyses bound GTP to GDP. Gα-GDP has no affinity for the effector protein and reassociates with the beta and gamma subunits.

5.5 CYTOPLASMIC SIGNALLING CASCADES

5.5.1 The MAP kinase cascade

Reference has previously been made to sequential protein kinase reactions involving phosphorylation and activation of multiple kinases in a signalling pathway, e.g. the ligand/receptor tyrosine kinase regulation of the mitogen-activated protein kinase (MAPK) family.

This section will seek to add further detail to this fascinating but complex picture. The general picture for the cytoplasmic signalling cascades, involving sequential protein phosphorylations, that originate in the cell membrane is summarized in Fig. 5.11.

The variety of signals that conscript the MAP kinase pathway demonstrates that this cascade serves a myriad of purposes and that the consequences of its activation will depend on cellular context.

Three distinct but related pathways or cascades, which have the general features described in Fig. 5.11, lead to the activation of three different sets of MAP kinases, which can translocate into the nucleus

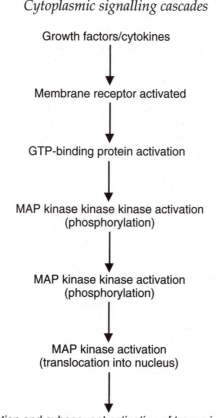

Growth factors/cytokines

Membrane receptor activated

GTP-binding protein activation

MAP kinase kinase kinase activation
(phosphorylation)

MAP kinase kinase activation
(phosphorylation)

MAP kinase activation
(translocation into nucleus)

Phosphorylation and subsequent activation of transcription factors

Fig. 5.11   The cytoplasmic signalling cascade.

where they selectively activate several different transcription factors (Su and Karin, 1996; Wang and Ron, 1996). Two of these pathways can participate in many overlapping stress responses, such as activation in response to inflammatory cytokines, treatment of cells with alkylating agents, ultraviolet light and ionizing radiation. These two pathways are consequently referred to as **stress pathways**. The other pathway that stimulates proliferation and differentiation will, for the purpose of this text, be referred to as the **mitogenic pathway**. This pathway can be activated in response to a wide variety of extracellular stimuli such as insulin, nerve growth factor (NGF), platelet-derived growth factor (PDGF), and in response to expression of the oncogenes, v-*src* and v-*ras*, in a cell-specific manner.

Activation of some transcription factors (Elk-1, ATF-2) can lead to cell proliferation (cell-cycle activation), differentiation and survival, while activation of others (c-Jun, GADD153) by the stress pathways mentioned above may lead to growth arrest and apoptosis. The relationships

Table 5.1   The mitogenic pathway

Generic name	Specific name
GTP-binding protein	Ras
MAP kinase kinase kinase	Raf
MAP kinase kinase	MEK1/2
MAP kinase	ERK1/2
Transcription factors	Elk-1, ATF-2

Table 5.2   The stress pathways

Generic name	Specific name	
	Stress pathway 1	Stress pathway 2
GTP-binding protein	Rac	?
MAP kinase kinase kinase	MEKK1	?
MAP kinase kinase	SEK1/MKK4*	MKK3
MAP kinase	JNKs/SAPKs	p38
Transcription factors	c-Jun, ATF-2	CHOP/GADD153

* also known as JNKK/SAPKK

between components within the three pathways are illustrated in Tables 5.1 and 5.2.

When receptor tyrosyl residues are phosphorylated as a result of ligand binding, adapter proteins that possess SH2 domains, such as Grb-2, can then bind and recruit guanine nucleotide exchange factors with proline-rich SH3 domains, e.g. m-Sos, to the membrane in proximity to Ras or Rac (Ras-related protein).

Ras and Rac are isoprenylated guanine nucleotide-binding proteins that behave as molecular switches: they are in the 'on' state when they bind GTP and are inactive when bound to GDP. The exchange factors promote the association of Ras and Rac with GTP, thus switching them to the 'on' state. The return from the 'on' state to the 'off' state is brought about by hydrolysis of the Ras-bound GTP to GDP. Ras has a slow GTPase activity, which can be increased at least tenfold by interaction with GTPase-activating proteins (GAPs). There are five Ras-specific mammalian GAPs known to date: these include p120^{GAP}, GAP1 and neurofibromin. The p120^{GAP} contains two SH2, one SH3 and a PH domain, together with a proline-rich region near to the N-terminal end.

Activation of the MAP kinases ERK1 and ERK2 (extracellular regulated kinases) is mediated by dual phosphorylations on threonyl and tyrosyl residues within the motif Thr185-Glu186-Tyr187 by the dual-specific MAP kinase kinase (MEK) which is itself activated by the Ser/Thr kinase activity of activated Raf (Fig. 5.12).

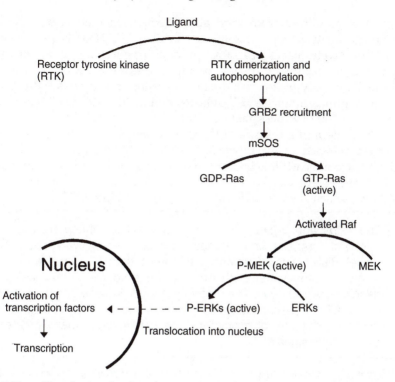

Fig. 5.12   Signal transduction from tyrosine kinase receptors. See text for details.

MEK lies at a point of convergence for multiple signals upstream, mediated by other protein kinases in addition to Raf, such as MEK kinase and Mos. Additional inputs that stimulate the MAP kinase pathway include the activation of protein kinase C (PKC).

The kinase domain of Raf is located at the C-terminus and is regulated by the N-terminal region. Activated Ras physically interacts with the N-terminal domain of Raf and thus abrogates the domain's regulatory influences. It is noteworthy that the oncogenic form of Raf has deletions in this N-terminal sequence, thus making it constitutively active. Activated Raf localizes to the plasma membrane. It is of interest, in this context, that both cytosolic and membrane-bound Raf are found in a large complex that includes two chaperones, Hsp90 and p50. Once Ras has done its job in localizing Raf to the membrane, it is no longer required to sustain Raf activity. Since both tyrosyl and threonyl phosphorylations are required to activate ERKs, phosphatases that remove phosphate from either site will inactivate them. Protein phosphatase 2A (PP2A) is the major phosphatase acting on MEK and ERK but only removes phosphate from seryl and threonyl residues. A new class of dual-specificity phosphatase known as CL100/C3H134 has been found that selectively inactivates ERKs by dephosphorylating both sites thus providing a negative feedback loop following MEK activation. Also, an immediate–early mitogen-inducible tyrosine phosphatase PAC1 has been identified in nuclei of T cells. Recombinant PAC1 *in vitro* is a dual-specificity Thr/Tyr phosphatase.

The pattern of a cascade of three activating kinases is also preserved in the stress-activated pathways. The kinase MEK kinase-1 (MEKK1) is distinct from the MEK activator Raf and does not lead to the activation of ERK1 and 2 but instead stimulates the stress-activated protein kinase (SAPK), which is identical to a c-Jun N-terminal kinase (JNK).

The v-*fos* and v-*jun* oncogenes were isolated independently as retroviral transforming genes carried respectively by the Finkel–Biskis–Jinkins murine osteogenic sarcoma virus and the avian sarcoma virus. Both are derived from the normal cellular genes c-*fos* and c-*jun*. The protein products of these genes are part of the AP-1 transcription factor. The AP-1 recognition sequence is found in a variety of promoters and can mediate many different extracellular signals.

Phosphorylation of c-Jun by JNK/SAPK occurs at five regulatory sites. Two of these sites are in the N-terminal transactivation domain (see below), which enhances transcriptional activity. The three other sites are clustered next to the C-terminal DNA-binding domain which inhibits DNA binding by c-Jun homodimers but has no measurable effect on DNA binding by c-Jun/c-Fos heterodimers (AP-1). The transcription factor ATF2 is also phosphorylated by JNK/SAPK on two threonyl residues.

The JNKs but not the ERKs are found to be persistently activated in apoptosis induced by gamma-radiation, UV-C and anti-Fas treatment. Overexpression of activated JNK causes cell death and a dominant-negative mutant of JNK prevents the UV-C and gamma-radiation-induced cell death.

The mechanisms by which JNK/SAPK and ERKs participate in cellular responses to mitogenic stimuli, environmental stresses and apoptotic agents are unclear. Binding of the Fas ligand and TNFα for example to their respective receptors results in stimulation of the enzyme sphingomyelinase, which catalyses the hydrolysis of cell membrane sphingomyelin to ceramide. Ceramide has been shown to be an activator of the JNK/SAPK cascade and an inhibitor of the ERK pathway (Verheij *et al.*, 1996). Stimulation of sphingosine kinase, on the other hand, with the concomitant increase in intracellular sphingosine-1-phosphate by, for instance, PDGF and protein kinase C (PKC) activation, leads to activation of the ERK-signalling pathway of cell growth and inhibition of the

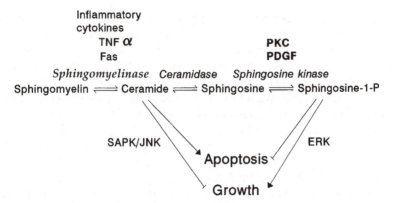

Fig. 5.13   Scheme depicting the effect of growth factors and cytokines on sphingolipid metabolism, leading to regulation of distinct members of the mitogen-activated protein kinase (MAPK) superfamily and culminating in cell growth or apoptosis. TNFα and Fas ligand stimulate sphingomyelin hydrolysis to form ceramide, an activator of JNK/SAPK and an inhibitor of ERK, whereas PDGF or activation of PKC increases sphingosine kinase activity and sphingosine-1-phosphate, a stimulator of the ERK-signalling pathway of cell growth and inhibitor of the JNK/SAPK-signalling pathway, leading to apoptosis. Source: redrawn from Cuvillier *et al.* (1996).

JNK/SAPK-signalling pathway. Thus, it is possible that cell survival/growth and apoptosis are dictated by a balance between ERK and JNK/SAPK signalling under certain circumstances.

The balance between ceramide-stimulated apoptosis and sphingosine-1-phosphate stimulated growth is depicted schematically in Fig. 5.13.

The binding of ligands such as PDGF, EGF, NGF and fibroblast growth factor (FGF) to their appropriate receptors, in addition to activation of kinase cascades, also induces the stimulation of phosphatidylinositol-4,5-bisphosphate (PIP_2) turnover by activating phospholipase C-γ-1 ($PLC-\gamma_1$) in a wide variety of cells. $PLC-\gamma_1$ contains SH2-binding domains that interact with phosphotyrosyl residues on the activated cell membrane receptors.

> Phospholipases C (PLCs) exist as multiple isoforms. Three types of PLC (b, g and d) can be distinguished on the basis of their molecular weights and amino acid sequences. Examples of these three types are $PLC-b_1$, $PLC-g_1$ and $PLC-d_1$, with molecular weights of 150, 145 and 85 kDa respectively.

In the FGF and NGF receptors, tyrosyl residues 1009 and 785 respectively bind $PLC-\gamma_1$ specifically, whereas, in the case of EGF receptors, $PLC-\gamma_1$ can bind to any one of five phosphorylated tyrosyl sites located in the C-terminal region of the receptor protein. The mechanism by which SH2 domains recognize distinct phosphotyrosyl-containing sites and the relevance of this selectivity for *in vivo* signalling is beginning to be unravelled. The PLC-γ_1 SH2 C-terminal domain has an extended hydrophobic cleft that binds +1 to +6 peptide residue C-terminal to the phosphotyrosine residue 1021. This is quite distinct from the Src SH2-binding pocket, and starts to address the mechanisms by which SH2-binding specificity is achieved. Site-directed mutagenesis has been used to investigate the binding specificity of the Src SH2 domain. By changing a single residue C-terminal to the phosphotyrosine, the specificity of the Src SH2 domain has been altered to resemble that of Grb-2. This mutant Src SH2 domain effectively substitutes for the Sem-5 (Grb-2 equivalent) in *C. elegans* in activation of the Ras pathway *in vivo*. Hence the binding specificity of an SH2 domain correlates with its biological activity.

The PH domains located near the N-terminal region of PLC-γ_1 may facilitate interaction between the phospholipase and the plasma membrane by possibly binding the membrane lipid PIP_2. The recruitment of PLC-γ_1 to membrane receptors (via SH2/phosphotyrosyl interaction) leads to tyrosine phosphorylation of PLC-γ and activation of phospholipase activity. PIP_2 is subsequently hydrolysed to diacylglycerol (DAG) and inositol-1,4,5-trisphosphate (IP_3), both of which serve as intracellular second messengers.

> PLC-b isoenzymes are activated by the GTP-bound alpha subunits of the Gq heterotrimeric G proteins. Receptors that use the Gqa/PLC-b route include thromboxane A2, bradykinin, bombesin, angiotensin II, histamine, vasopressin, cholinergic muscarinic receptors and thyroid-stimulating hormone.

Protein kinase C (PKC) is activated by DAG in the presence of Ca^{2+}, and IP_3 stimulates the release of Ca^{2+} from intracellular Ca^{2+} stores, leading to increased transients of cytosolic Ca^{2+}. The smooth endoplasmic reticulum (ER), which is a continuous network throughout the cell, is thought to be the major Ca^{2+} store from which IP_3 stimulates this Ca^{2+} release. IP_3 receptors are numerous in the smooth ER but scarce on the rough ER.

The release of Ca^{2+} from the smooth ER stimulates the further release of Ca^{2+} from calcium-sensitive calcium pools leading to the generation of cytosolic Ca^{2+} waves which can propagate into the nucleus. A rise in nuclear Ca^{2+} concentration affects nuclear functions such as cell-cycle regulation, gene transcription and the induction of apoptosis (Fig. 5.14).

The induction of high levels of protein synthesis is an essential step in triggering cellular proliferation. The PDGF receptor, for example, possesses an autophosphorylation site at tyrosine 751 that is capable of recruiting the p70S6 kinase (p70S6k). The activation of p70S6k is independent of the Ras pathway since it bifurcates from Ras at the PDGF receptor. The activated p70S6 kinase is capable of effecting multiple phosphorylations of the 40S ribosomal protein S6.

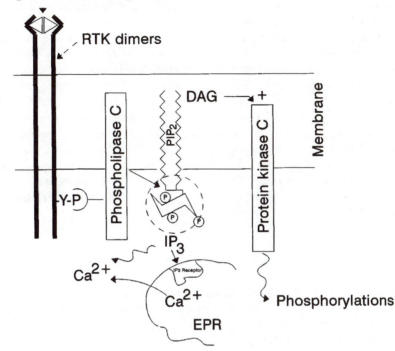

Fig. 5.14 Recruitment and activation of phospholipase C to autophosphoryl-ated ligand-bound PDGF receptor. Activation of phospholipase C leads to cleavage of the membrane phospholipid phosphatidylinositol-4,5-bisphosphate (PIP_2) into inositol trisphosphate (IP_3) and diacylglycerol (DAG). IP_3 promotes release of Ca^{2+} from storage organelles. Diacylglycerol stimulates protein kinase C, which phosphorylates certain serine- or threonine-containing proteins.

The 40S ribosomal subunit that contains phosphorylated S6 may selectively recognize polypyrimidine tracts located immediately after the N^7GTP cap in a family of mRNAs that include transcripts for the manufacture of ribosomal proteins, elongation factors and other proteins of unknown identity.

Receptor tyrosine kinases are capable of transmitting signals for differentiation in some cells while signalling cell proliferation in others although both signalling systems work through Ras. It is possible that distinguishing between differentiation and proliferation could be achieved by different targets for Ras and Raf in different cell types. There are, however, several instances where differentiation and proliferation signals are mediated by Ras and Raf signalling via the MAP kinases ERK1 and ERK2.

In PC12 phaeochromocytoma cells, activation of Ras by EGF causes proliferation but activation of Ras by NGF induces differentiation. Since both EGF and NGF signal through ERKs, the question is, how do these two signals lead to two different outcomes?

Quantitative differences in Ras-mediated ERK activation are thought to be responsible for the different decisions made by the cell. Gradients of extracellular factors are important during development and may give rise to the quantitative differences that can induce developmental responses in apparently identical cells. Sustained activity of ERKs in response to NGF in PC12 cells leads to ERK translocation to the nucleus and cell differentiation while EGF leads to transient ERK activation and cell proliferation. Fine tuning of duration of ERK activity may be achieved by ERK-specific phosphatases. The location of these phosphatases may also be important. For example, if they are located in the nucleus they could determine the cytoplasmic ERK activity relative to nuclear activity.

5.6 CONCLUSION

The constellation of protein kinases that are activated as a result of ligands interacting with cell membrane receptors determines the final biological response to the cell surface signals. The existence of, for instance, cascades of kinases with branching interactions:

• increases the number of steps at which the original signals can be diversified;

• is consistent with the fact that downstream targets usually require two or more inputs for functional regulation;

• allows diversification of the signal in order to establish cross-checks to ensure that all systems are 'go' before a commitment is made to alter a functional readout, especially if the latter determines cell fate.

B. THE INTERPLAY BETWEEN CYTOTOXIC CELLS AND CYTOKINES IN THE REGULATION OF CELL DEATH

5.7 NATURAL KILLER CELLS AND CYTOTOXIC T LYMPHOCYTE KILLING MECHANISMS

5.7.1 Introduction

In addition to the antibody response to foreign antigens, the immune system has developed a cell-mediated killing response to counteract the intracellular threat posed by viruses, certain bacteria and tumour cells. Key players in the cell-mediated response are cytotoxic T lymphocytes

(CTLs), natural killer (NK) cells, lymphokine-activated killer (LAK) cells and tumour-infiltrating lymphocytes (TILs).

CTLs recognize cells presenting pathogen-derived antigenic peptides on major histocompatability complex (MHC) class I molecules, while NK cells target cells that do not have the MHC class I molecules (for further details on these cells and their activation see Bowen and Bowen, 1990). Recently, it has been established that NK cells do recognize either the absence or major changes in MHC class I antigens and that this recognition represses rather than activates NK cells. (For a review on this topic, see Lanier and Phillips, 1996.) In fact, NK cells and CTL cells are able to regulate each other's function (Kos and Engleman, 1996). In many viral infections NK cells are the first to respond, followed by CD8$^+$ CTLs, which first complement the reaction of NK cells and then replace their activity. (CD8$^+$ is a transmembrane protein with an extracellular domain. It is expressed on CTL and, like T-cell receptors, recognizes MHC proteins. Unlike the receptors, it binds to the non-variable portion of the MHC class I molecule.) Kos and Engleman (1996) propose that NK cells are involved in the differentiation of CD8$^+$ CTLs and that these then downregulate NK-cell activity. CTLs produce interleukin 4 (IL-4) and transforming growth factor-beta (TGF-β), which inhibit NK cells. Tumour-infiltrating lymphocytes (TILs), which can be isolated from tumour specimens, show specific MHC-restricted killing of autologous tumour cells and are generally found with the CD8$^+$ subset of T cells. These antigen-specific TIL have been more effective in killing, when tested in animal tumour models, than the antigen-non-specific LAK cells. Like LAKs they are activated by lymphokines such as IL-2 but, unlike LAKs, they have the ability to traffic (move) to tumour sites.

Cytotoxic lymphocytes appear to be able to use two independent mechanisms for triggering cell death in their target cells. One of these pathways involves secretion of granules containing perforin and other cytotoxic substances and is called the granule exocytosis mode of killing; this mechanism is also used by NK cells. The second pathway, which involves interaction with Fas ligand, is non-secretory and is not used by NK cells. There is some evidence (Vujanovic *et al.*, 1996) that NK cells also have non-secretory mechanisms for inducing apoptosis and that these may involve cell-membrane-bound ligands of the TNF family.

5.7.2 Perforin-mediated killing by CTL and NK cells

The main pathway for killing by CTL, NK and LAK cells is mediated through perforin, a pore-forming protein (Fig. 5.15).

The structure and function of perforin was reviewed by Liu, Walsh and Young (1995). Perforin is a glycoprotein (534 amino acids) and has sequence homology to complement component C9. Like C9, perforin is able to integrate into cell membranes, aggregating to form polyperforin

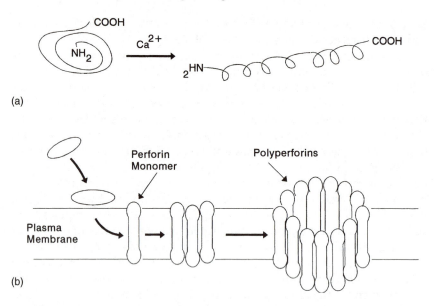

Fig. 5.15 A schematic representation of the effect of Ca^{2+} in promoting conformational changes in perforin monomers, which allows them to form cylindrical structures. These then insert into the plasma membrane of the target cell, forming polyperforin pores. Source: redrawn from Liu, Walsh and Young (1995).

pores made of between 12 and 18 perforin monomers (Shinkai, Takio and Okumura, 1988; Lichtenheld *et al.*, 1988). Perforin is found in most NK cells and mainly in CD8$^+$ CTL cells and $\gamma\delta$T cells (Nakata *et al.*, 1990). ($\gamma\delta$ is a complex of two T-cell surface receptors that helps to transduce the signal when antigen MHC complex binds to T-cell receptors.) Several lymphokines can rapidly induce perforin mRNA in CD8$^+$ cells, e.g. interleukin 2 (IL-2), interleukin 6 (IL-6; Smyth *et al.*, 1991) and interleukin 12 (IL-12; Salcedo *et al.*, 1993). TGF-β, on the other hand, inhibits perforin stimulated by IL-2 and IL-6 (Smyth *et al.*, 1991). These cytokines and their effects on NK and CTL killing, as well as their roles in promoting or preventing apoptosis, are further discussed at the end of this chapter.

Perforin appears to be crucial for the granule exocytosis mode of cell killing. When the NK, LAK and CTL cells interact with their target cells they release the contents of their cytoplasmic granules into the space between the killer and target cell (see Fig. 5.16). These granules contain perforin, which, in the presence of Ca^{2+} (which is essential for polymerization), forms polyperforin pores in target cell membrane (Fig. 5.15; Liu, Walsh and Young, 1995).

Perforin, although capable of lysing cells on its own, does not initiate the DNA fragmentation characteristic of CTL and NK killing; instead it causes loss of volume control by punching holes in the target cell membrane, resulting in classical necrosis. The exocytosed granules contain, in addition to perforin, a variety of cell-death effector molecules such as TNF (Liu *et al.*, 1987), a polyadenylate binding protein T1A-1 (Tian *et al.*, 1991), secreted ATP (Di Virgilio *et al.*, 1990), leukalexin (Liu *et al.*, 1987) and serine esterases called granzymes (Redelman and Hudig, 1980; Lobe *et al.*, 1986; Hudig, Ewoldt and Woodward, 1993), also named fragmentin by Shi *et al.* (1992). It is a combination of perforin and these effector molecules that induces lysis with fragmentation of target DNA (apoptosis), which is typical of CTL and NK killing.

Details of how all these death-promoting factors initiate apoptosis are vague, but TNF can trigger apoptosis by activating the TNFR (section 5.4.3).

Much work has focused on one of the granzymes, granzyme B, which has been implicated in the activation of ICE-like proteases and the cleavage of PARP (section 4.8.5; Quan *et al.*, 1996). Froelich *et al.* (1996) have shown that exposure of Jurkat cells to granzyme B and perforin results in cleavage of PARP to an apoptotic 89 kDa fragment and to lesser amounts of a 64 kDa fragment. The 64 kDa fragment was produced directly by granzyme B. Intracellular granzyme B appears to be translocated to the nucleus, where it directly cleaves PARP (Pinkoski *et al.*, 1996; Trapani *et al.*, 1996). Presumably, the 89 kDa fragment was cleaved by an ICE-like protease cascade initiated by granzyme B. Several caspases, including caspase-1, caspase-3, caspase-8 and caspase-10, have been shown to be processed to their active forms by cells undergoing apoptosis induced by granzyme B (Darmon, Nicholson and Bleackley, 1995; Shi *et al.*, 1996a).

Interestingly, granzyme B can induce apoptosis in target cells at any stage of the cell cycle, but in order for cell death to occur cyclin-A-associated cyclin-dependent kinase must be activated (Shi *et al.*, 1996b). DNA fragmentation and nuclear collapse seen in granzyme-B-induced apoptosis can be prevented by inactivating cyclin-dependent kinase (Shi *et al.*, 1994). Consequently CTL, LAK and NK cells are able to kill all cycling cells targeted. This is consistent with results obtained by Khalil *et al.* (1990) on NK cells and Nishioka and Welsh (1994) on CTLs, where reduced susceptibility to NK and CTL killing was shown if target cells were in G_0. Non-cycling cells do not express cyclin-dependent kinase (McGowan, Russell and Reed, 1990) and this enzyme appears to be essential for granzyme-B-mediated cell death. Shi *et al.* (1996b) indicate that, when cyclin-bound Cdc-2 and Cdk-2 kinases are activated during granzyme apoptosis, Wee-1 kinase (section 3.8) can rescue cells from granzyme-induced apoptosis by preventing Cdc-2 dephosphorylation (Chen *et al.*, 1995).

5.7.3 Non-secretory CTL killing

Cytotoxic lymphocytes devoid of the perforin-secreting mechanism for

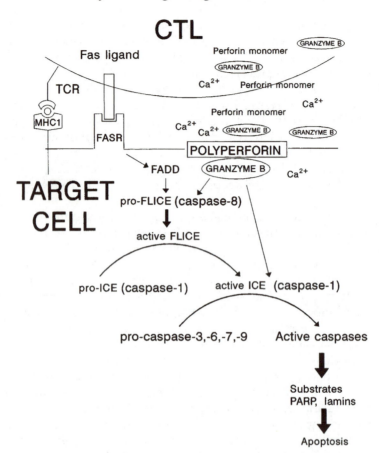

Fig. 5.16   Schematic figure depicting the granule exocytosis and non-secretory modes of killing by cytotoxic T lymphocytes. In order to initiate either of the above killing responses, CTL cells must recognize target cells presenting antigenic peptides on MHC1 molecules. This initiates the accumulation of cytoplasmic granules containing perforin and granzymes (including granzyme B), plus other cytotoxic molecules (not represented in the figure), at the interface between the CTL and the target cell. The granules are then exocytosed into the extracellular space, where the elevated Ca^{2+} ion concentration induces conformational changes in the perforin molecules depicted in Fig. 5.13. This allows the granzymes and other cytotoxic molecules to enter the target cell through the polyperforin pore. Granzyme B can then initiate the conversion of procaspases to their active form, leading to the breakdown of PARP and lamins, and thus triggering apoptosis. The Fas-dependent pathway is initiated by upregulation of Fas ligand expression on the CTL cell. Binding with Fas-receptor molecules (FASR) on the target cell leads to the induction of apoptosis. The cell surface receptor molecule, when activated, uses an adapter molecule, FADD, to physically engage a cytosolic apoptotic protease pro-FLICE (caspase-8), converting it to its active form and initiating a cascade of caspases in a similar fashion to granzyme B. Hence both the granule exocytosis (secretory) and the non-secretory modes of CTL killing converge at the ICE-like proteases, leading to the activation of similar caspases and apoptotic death.

cell killing are still able to induce apoptosis if the target cell expresses Fas ligand in its membrane (Walsh *et al.*, 1994). Fas ligand is expressed on the cytotoxic T lymphocyte membrane (Suda *et al.*, 1993), allowing CTL cells to induce Fas/Fas-ligand cell killing (Fig. 5.16).

Cell death initiated by the interactions of various ligands with apoptotic-inducing cell surface receptors has been demonstrated in a variety of cells. Krammer *et al.* (1994) have reviewed ligand triggering of apoptosis in the immune system.

The non-secretory cytocidal activity of CTL cells was indistinguishable from that induced by the secretion of granules containing perforin and their cytotoxic substances (e.g. granzyme B; Walsh *et al.*, 1994). This is not surprising in view of the discovery that the Fas/Fas ligand pathway, through activation of the FADD death domain, activates FLICE (caspase-8; section 4.8.7), which in turn, like granzyme B found in secretory granules, is able to induce a cascade of caspases (ICE-like proteases) that leads to the breakdown of PARP and lamin A (section 4.8.5). Thus, the two pathways for cell killing converge around the ICE-like proteases, triggering the proteolysis required to convert the procaspases into active caspases (Fig. 5.16).

NK and CTL mechanisms of killing have recently been reviewed by Kagi *et al.* (1996) and Berke (1997).

5.8 CYTOKINES AND THEIR ROLE IN PREVENTING AND PROMOTING APOPTOSIS

5.8.1 Introduction

Several of the cytokines called interleukins, which are factors produced and released by activated T lymphocytes that act on other lymphocytes to produce biological effects, can prevent or rescue cells from apoptosis. They include interleukin 2 (IL-2), interleukin 3 (IL-3), interleukin 4 (IL-4), interleukin 7 (IL-7), interleukin 9 (IL-9) and interleukin 13 (IL-13). This inhibition of cell death is usually associated with increased levels of Bcl-2 or Bcl-X_L protein in the target cells. Cytokines, as stated earlier in this chapter, function in a redundant manner as they often activate the same transduction mechanisms; others have different effects depending on the presence or absence of other cytokines. One example is IL-13's ability to inhibit IL-2-induced proliferation (Chaouchi *et al.*, 1996). Interleukin 4 may interfere with the stimulatory effects of IL-2 on lymphokine-activated killer cells and gamma-interferon can in turn inhibit the response of B lymphocytes to IL-4. Interleukin 9, a major anti-apoptotic factor for thymic lymphomas, has also been reported to stimulate protease expression in mast cells and T-helper cells. One of these proteases is granzyme B, which is strongly associated with apoptotic cell death (section 5.7.2; Louahead *et al.*, 1995).

5.8.2 IL-3, IL-4, IL-5

Several cytokines produced by T lymphocytes are involved in B-cell development. IL-4 stimulates the early activation of resting B cells, while IL-5 promotes their proliferation and IL-6 induces the final maturation to immunoglobulin secretory cells. Interleukin 13 acts at different stages of B-cell maturation with a spectrum of biological activities overlapping those of IL-4. Interleukin 3 is a member of a family of growth and differentiation glycoprotein regulators collectively called colony-stimulating factors (CSF) and strictly speaking is not an interleukin, as it targets a range of non-lymphocyte cells and interleukins are defined as cellular signalling agents between lymphocytes. Interleukin 3 has a broad spectrum of biological activities, stimulating proliferative responses in cells of both erythroid and myeloid lineage, suggesting that it regulates the proliferation and differentiation of early haemopoietic multipotential cells in marrow. Interleukin 3 can also promote proliferation of precursors of erythroid cells, megakaryocytes, eosinophils, mast cells, neutrophils and macrophages. This accounts for its multiplicity of names given to IL-3 before it was sequenced. Withdrawal of such growth factors as CSF, IL-3 and IL-4 not only causes cessation of cell population growth but eventually the cells undergo apoptosis (see Fig. 4.7)

Prevention of apoptosis in haemopoietic cells by IL-3 and IL-4 requires phosphatidylinositol-3-kinase (PI3-kinase) activity (section 5.5; Minshall *et al.*, 1996). Insulin-like growth factor (IGF) — which, as mentioned in section 4.8.6, inhibits death by preventing activation of the protease mediators of apoptosis and increasing the activity of anti-apoptotic proteins — and nerve growth factor also require the activity of PI3-kinase in order to prevent apoptotic cell death (Yao and Cooper, 1995). Yet IL-5 and granulocyte macrophage colony-stimulating factor (GM-CSF) are able to ensure survival even if PI3-kinase has been inhibited by wortmannin (section 5.4), which is a potent inhibitor of this kinase (Arcaro and Wymann, 1993). Thus, phosphatidylinositol-3-kinase is important in signalling the inhibition of apoptosis in haemopoietic cells and the nervous system and may be generally used to prevent apoptosis; however, other signalling mechanisms for preventing cell death must also exist, since this kinase can be bypassed by IL-5 and GM-CSF.

5.8.3 IL-6

Interleukin 6, in addition to being produced by T lymphocytes, is also synthesized by B cells, monocytes, fibroblasts, endothelial cells, keratinocytes, astrocytes, mesangial cells and bone marrow stroma cells. A diverse group of molecules can induce its production, including lipopolysaccharide, IL-1, TNF and dsRNA viruses. Although IL-6 was originally characterized as a cytokine that induces the final maturation of B cells to immunoglobulin-secreting cells, it has a bewildering array of functions. Interleukin 6 acts synergistically with IL-3 to support the formation of multilineage blast cell colonies and induces not only proliferation but also the differentiation of CTL cells in the presence of IL-2. Interleukin 6 is a major growth factor for malignant plasma cells (myeloma cells) and is directly implicated in the pathogenesis of this and other malignant diseases. The biology of IL-6 has been reviewed by Kishimoto (1989), who discusses the role of IL-6 in disease.

Interleukin 6 is essential for the survival of peripheral blood early plasma cells, as well as their differentiation into mature plasma cells in the bone marrow, and it is able to delay the onset of apoptosis in these early plasma cells. Interleukin 6 is reported to inhibit apoptosis in myeloma cells (Lichtenstein *et al.*, 1995), thus contributing to the expansion of myeloma clones. The mechanism of inhibition does not involve Bcl-2 expression, as it is actually decreased in the presence of IL-6; however, IL-6 upregulates Bcl-X_L protein (section 4.5.1), which then blocks apoptosis in these cells (Schwarze and Huwley, 1995). Interestingly, Rensing-Ehl *et al.* (1995) have published evidence that Fas/Apo-1 can, in addition to activating the cell-death transduction mechanism, also activate genes via NF-κB, inducing IL-6 production. This implicates Fas in a broader range of biological functions than just cell death. IL-6, like IL-2 and IL-12, can induce perforin production in NK and CTL cells, thus contributing to the role of these cells by acting as a lymphokine activator for NK and CTL killing. Thus, although IL-6 is undoubtedly involved in promoting myelomas, it also contributes to the surveillance of malignancies through activation of the two major cell types that help to ensure that tumour cells do not survive.

5.8.4 IL-2

Interleukin 2 is a central mediator of the growth activity of both B and T cells and also cytotoxic cells, upgrading their activity and converting NK and CTL cells into LAKs. The ability to stimulate continuous growth of primary T-cell lines in culture is the major distinguishing property of IL-2 and led to its discovery by Morgan, Ruscetti and Gallo (1976). Withdrawal of IL-2 induces apoptosis in these cells. There are conflicting reports as to which of the Bcl protein survival factors declines when IL-2 is withdrawn. Broome *et al.* (1995) have demonstrated that it is a dramatic decline in Bcl-X_L and not Bcl-2 protein levels that precedes IL-2-induced apoptosis. On the other hand, antigen-specific memory T cells, which have low levels of Bcl-2 (Akbar *et al.*, 1993), are apoptosis-prone and can be rescued from radiation- or dexamethasone-induced apoptosis by IL-2; their survival correlates well with the induction of Bcl-2 in these cells (Mor and Cohen, 1996). Akbar *et al.* (1996) report that both *bcl-2* and *bcl-X$_L$* genes are induced by IL-2 rescue of T cells, but there is little change in the expression of the apoptotic promoter genes *bax* and *bcl-X$_S$*. Interleukin 2, like IL-12 and IL-6, rapidly induces perforin mRNA in CD8$^+$ CTL cells and NK cells, activating them to lymphokine activator killer cells and increasing their effectiveness in cell killing. Interleukin 2, in addition to augmenting this cytolytic activity of NK and CTL cells, also increases their cytokine secretion and their proliferation. Interleukin 2 is consequently essential in preventing apoptotic cell death in T cells and also increases the apoptotic potential of cytotoxic cells.

5.8.5 IL-12

Interleukin 12 enhances gamma-interferon (IFN-γ) production by NK cells and T lymphocytes and plays a pivotal role in the development of Th$_1$ cells, which secrete IL-2 and IFN-γ. Interleukin 12 prevents the differentiation of T cells into Th$_2$ cells, which secrete IL-4 and IL-10. (Th$_1$ cells are helper T cells that activate macrophages to destroy micro-organisms they have ingested; Th$_2$ cells are helper T cells that stimulate B cells to proliferate and secrete antibodies.) Jeannin *et al.* (1996), however, have demonstrated that IL-12 can synergize with IL-2 and other stimuli to induce IL-10 and IL-4 production by T cells, confirming that a complex cytokine network is involved in the induction and duration of Th$_1$ and Th$_2$ T-cell responses. Interleukin 12 induces tyrosine phosphorylation of Tak2 and Tyk2, while IL-2 induces phosphorylation of Jak1 and Jak3 in NK cells, yet both increase the lytic activity of NK cells (Yu *et al.*, 1996). In fact, IL-12 is an extremely potent inducer of NK proliferation and NK production of IFN-γ and this helps to explain its effectiveness in tumour regression. It is even effective in animals bearing fairly large tumours that do not respond to other cytokines, such as IL-2 (Brunda *et al.*, 1994). Tannenbaum *et al.* (1996) report that daily treatment with IL-12 provided almost complete tumour regression, while in untreated mice the tumours continued to grow. The tumour regression seen in IL-12-treated mice is associated with the enhanced expression of perforin and granzyme B in NK cells and thus increases the NK-induced apoptosis of the tumour cells. This cytokine may also prove its efficacy in the treatment of human cancers.

5.8.6 IL-10

Conflicting reports can be found in the literature about the role of IL-10 in cell death. It has been reported as preventing apoptosis of germinal centre B cells (Levy and Bronet, 1994) but induces apoptosis in B chronic lymphocytic leukemia cells (Fluckiger, Durand and Bauchereau, 1994); this promotion or prevention is again associated with the expression of

Bcl-2 protein. Interleukin 10 has also been reported as partially inhibiting lipopolysaccharide (LPS)-induced apoptosis, which is mediated through the action of TNF (Eissner *et al.*, 1995). This is not too surprising as IL-10 is an antagonist of the pro-inflammatory cytokines, such as TNF and IFN-γ (Moore *et al.*, 1993; Gerard *et al.*, 1993).

5.8.7 IFN-γ

IFN-γ and TNFα suppress both early and late stages of haemopoiesis by induction of apoptosis (Selleri *et al.*, 1995). These two cytokines have been demonstrated, in the absence of IL-2 and IL-4 secretion by peripheral blood mononuclear cells (PBMCs), to upregulate Fas antigen on haemopoietic cells (Oyaizu *et al.*, 1994). Nagafuji *et al.* (1995) have demonstrated that CD34+ (haemopoietic progenitor) cells, freshly isolated from bone marrow, that did not express Fas, were induced by IFN-γ and/or TNFα to express Fas, in a dose-dependent fashion, on their surface. The two cytokines together have a synergistic effect on Fas induction in these cells. Interestingly, TNFR1, which mediates TNFα-induced cytotoxicity, also mediates the IFN-γ-induced Fas expression. The induction of Fas expression on haemopoietic progenitor cells could thus help prevent uncontrolled proliferation.

5.8.8 TNF

Tumour necrosis factor (TNF) was named by Carswell *et al.* (1975) for its massive lethal effect on certain tumours. Two forms of TNF have been identified: TNFα and TNFβ (lymphotoxin). Urban *et al.* (1986) reported that TNF was a potent effector molecule for tumour cell death, mediated via activated macrophages. This tumour-related cell death can take two forms: haemorrhagic necrosis involving the inflammatory role of TNF (section 6.9.2) and apoptosis. Among the activities of this multifunctional cytokine are stimulation of fibroblast growth, T-cell proliferation, involvement in inflammation and the inhibition of lipoprotein lipase, which leads to cachexia (a general loss of body weight). TNFα has also been called cachectin (Beutler and Cerami, 1986) because of the wasting (loss of body weight) so often associated with malignant tumours. TNFα exists as a membrane-bound precursor molecule of 26 kDa that is processed to the 17 kDa mature form by TNFα-converting enzyme (TACE), which is a membrane-bound metalloproteinase (Moss *et al.*, 1997). TNFα can either remain membrane-bound or be secreted. Shih and Stutman (1996) have reported that target cells respond to the TNF cytotoxic signal, mainly at the G_1–S boundary, and begin to die apoptotically as they exit from the S phase. TNF exercises its apototic effect by activating TNFR (section 5.4.3) and is related to the Fas/Apo-1 mechanism. The dramatic effects of this cytokine on tumours led to many attempts to use it as an anti-tumour therapeutic agent in the late 1980s and early 1990s. Systemic administration of TNF in tumour therapy has proved disappointing, as it has only limited effects on tumours while being highly toxic to humans.

Jäättela *et al.* (1996) demonstrated that A20 (which is a zinc finger protein, the product of a cytokine-induced primary response gene) effectively inhibits signalling from TNF and IL-1 receptors. The interference with cytokine signalling appears to be at an early stage, after ligand binding to the receptor molecules and before the activation of second messenger pathways. A20 can protect cells from TNF-mediated cell death (Opipari *et al.*, 1992) but does not protect cells induced by Fas or LAK cells. Other proteins that inhibit TNF-induced cell death are Bcl-2, Bcl-X$_L$ and CrmA (Jäättela *et al.*, 1995; Tewari and Dixit, 1995), but these can also inhibit Fas-induced death. The mechanism of protection by A20 appears to be different from that of the other proteins and none of the other proteins interfere with TNF NF-κB activation. Jäättella *et al.* (1996) suggest that TRADD (TNF-receptor-associated protein), which signals both cell death and NF-κB activation (Hsu, Xiong and Goeddel, 1995; see Fig. 5.12) may be the target for A20, interfering with TRADD binding. However, this would not explain how A20 inhibits signalling from the IL-1 receptor, which does not bind to TRADD.

5.8.9 IL-1

Several cell types have the capacity to produce IL-1, but its release is seen mainly in monocytes/macrophages and is often associated with the apoptosis of the producer cells. Lipopolysaccharide (LPS)-activated monocytes increase their secretion of IL-1β, which is a major inflammatory cytokine and also stimulates the immune response, inducing the synthesis of other cytokines. Interleukin 1β has been implicated in the inhibition of Fas-mediated apoptosis in synovial cells (Tsuboi *et al.*, 1996), with Fas antigen being downregulated after treatment with IL-1β.

5.8.10 TGF-β

Tumour growth factor-beta (TGF-β$_1$) administered exogenously to unstimulated NK cells, inhibits NK cell proliferation and production of the cytokines TNFα, IFN-γ and GM-CSF, and downregulates NK cytotoxicity (Bellone *et al.*, 1995). These effects of TGF-β$_1$ on NK cells can be overcome by NK-activating cytokines IL-2 and IL-12; thus NK activation is determined by a balance of inhibitory or stimulatory cytokines. The inhibition of cell-cycle progression of NK cells by TGF-β probably occurs by inhibitory phosphorylation of the retinoblastoma protein, through modulation of CDK inhibitors such as p15 (see Fig. 3.14). TGF-β has been reported by Arteaga *et al.* (1993) to promote a human TGF-β-producing breast cancer cell line, probably by downregulating NK cells and their apoptosis-inducing cytokines. Antibodies to TGF-β restored NK cell activity and tumour formation was blocked.

TGF-β has been reported to prevent cell-cycle progression in a number of cell types and is associated with increased apoptosis. However, the effects of this multifunctional cytokine depend on the type of cell, its state of differentiation and the presence of other cytokines, such as IL-2. For example, Ayoub and Yang (1996) showed that TGF-β_1 inhibited stationary cultures of IL-2-dependent CD4 bovine lymphoblastoid T cells (BLTC) by downregulating their IL-2 receptor (IL-2R) expression, arresting cells in G_0/G_1 of the cell cycle and inducing apoptosis in these cells. Small quantities of IL-2 reversed these effects, with TGF-β_1 augmenting the proliferative response of BLTC through up-regulation of IL-2R expression, allowing cell-cycle progression and preventing apoptosis. The effects of TGF-β_1 alone are the opposite of those seen when TGF-β_1 and IL-2 are present. Other workers (Zhang *et al.*, 1995) have also reported that TGF-β_1 and IL-2 synergize to prevent apoptosis and promote cell expansion in CD4 T effector cells. Since these effector cells are specialized to help B cells develop into antibody-secreting plasma cells, these results suggest that the availability of IL-2 and TGF-β, as well as antigens, is probably an important factor influencing the size and duration of the primary antibody response.

5.9 CONCLUSION

The second part of this chapter attempted to provide the reader with an insight into how cytotoxic cells of the immune system trigger apoptosis and to relate this to some of the key molecules already discussed in Chapter 4. The reader was also introduced to some of the cytokines that influence whether a cell lives or dies. The information in section 5.8 is not comprehensive, but it does attempt to provide a base to build upon.

CHAPTER 6

Cell death in tissues and tumours

6.1 INTRODUCTION

Previous chapters have focussed on the definitions, mechanisms and control of mitosis and apoptosis. This final chapter will look in greater detail at the incidence of apoptosis in normal and abnormal development and the tissue kinetics of cell loss. The first part of the chapter will deal briefly with tissue kinetics, aspects of cell death in the development of animals and plants, the thymus, immunology, haemopoietic tissues and intestine. The second and concluding section will give an outline of cell death in tumours, an increasingly important topic.

Self-renewing systems such as the intestine and bone marrow maintain a more or less constant population of cells, where cell loss through apoptosis equals cell production through mitosis. In growing tumours there is an overall gain in cell number, but even here cells are constantly being lost from the population by apoptosis, necrosis, migration and exfoliation. It is the aim of tissue kinetics to define each cell population in terms of its size and change of size (or flux) with time. The discipline has essentially been outlined and introduced by Aherne, Camplejohn and Wright (1977).

6.2 THE TISSUE KINETICS OF CELL LOSS

The tissue kinetics of cell loss were usefully reviewed by Wright (1981). He indicated the early emphasis placed on measurement of cell proliferation. Even so, Steel (1966, 1968, 1977), drawing on data obtained from experimental tumours, was able to calculate the contribution made by cell loss. He showed that the growth rate of a cell population, taken as the time required for the number of cells in the population to double, was seldom equal to the birth rate as reflected by the cell-cycle time. This revealed that a number of cells were resting, differentiating, migrating and/or dying, leading to a measurable level of overall cell loss. Such calculations enabled subsequent studies on the kinetic responses of tumours to irradiation and chemotherapy.

One other important consideration when measuring cell death or cell loss is whether we are talking of actual cell removal or rather loss of reproductive viability. This again is particularly important in relation to tumours, since the ability of a tumour to regenerate after irradiation or chemotherapy depends more on the survival of cells with reproductive capacity (so-called clonogenic cells) than on the overall number of cells destroyed by the treatment.

Cell death in tissues and tumours often appears to be time- or age-dependent. Cells are born, live for a given period of time and then die. This life span varies from tissue to tissue. In most tissues, cells are continually leaving the cell cycle (Chapter 3) and most, unlike the so-called dividing or stem cells, lose their reproductive capacity and differentiate, specializing in a particular form and function before death. Cells can also leave the cycle to enter a resting stage called G_0, from which they can re-enter the proliferative cycle if conditions are appropriate. Resting cells appear either to be able to leave or enter the cell cycle before the onset of DNA synthesis at G_1 phase or, in the case of epidermis and liver, to decycle in G_2 phase.

Although the organization of tissues may be complex, they can kinetically be divided into two compartments: proliferative (P) and non-proliferative or quiescent (Q). The former compartment contains mitotically competent cells while the latter includes decycled, resting, differentiating and dying cells. Wright (1981) has extended this concept to consider operational behaviour in tissues generally and describes a three-compartment cell kinetic model (Fig. 6.1). Using this model and the work of Steel (1968), he goes on to calculate the rate of cell loss and death in a range of tissues and tumours.

In a growing population of cells where there is theoretically no cell loss, population size is a function of the birth rate. Thus, birth rate k_B will be equal to growth rate k_G.

If, as is more normal, cell loss is occurring, then k_G will be less than k_B:

$$k_L = k_B - k_G$$

where k_L is the cell loss rate expressed as cells/1000 cells per hour.

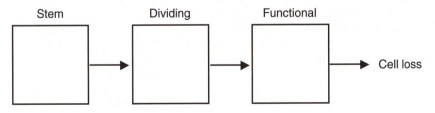

Fig. 6.1 A simple three-compartment model showing how a renewing cell population can be classified according to kinetic behaviour. Source: redrawn from Wright, 1981.

Steel (1968) introduced a cell loss factor, θ, which expressed the rate of cell loss compared with the rate of birth, assessing to what extent cell loss was affecting the proliferative potential of a cell population:

$$\theta = k_L / k_B$$

A similar calculation of apoptotic rate was given by Sarraf and Bowen (1986), and a very useful current review of the tissue kinetics of tumours is presented by Alison and Sarraf (1997).

6.3 CELL DEATH IN EMBRYOGENESIS AND DEVELOPMENT

Both animal and plant development are invariably associated with growth. The significant involvement of cell death therefore appears at first glance somewhat paradoxical. It is, however, clear that substantial levels of cell death not only occur during normal embryological or larval development as an integral part of the morphogenetic or tissue-moulding process, but also persist throughout life, maintaining normal tissue balance and kinetics. This all-important kinetic balance may be perturbed by states of disease and ageing and such conditions often provide us with further insight into factors controlling cell death. This means that the potential number of examples that could be drawn on is almost infinite, in that virtually all multicellular living things are as much a product of differential cell death as they are products of differential cell division. Areas of significant interest that have been reported in the recent literature include examples taken from vertebrate development, such as the shaping and development of the brain and spinal ganglia, morphogenesis of the limbs and interdigital cell death, the development of the thymus, haemopoetic system, heart, digestive tract, reproductive system, prostate, uterine epithelium, secondary palate, optic cup, retina and lens, and amphibian metamorphosis. A similarly wide range of developmental cell-death studies can be found in invertebrates, the most significant dealing with nematode and insect development, where definitive suicide genes have been identified (Chapter 4). Programmed cell death in insect and amphibian metamorphosis provides interesting and illuminating comparisons. Some areas of programmed cell death during the development of animals have been reviewed by Milligan and Schwartz (1996). A few of the major examples will be examined here.

6.3.1 Cell death and limb development

The role of programmed cell death in limb development was extensively reviewed by Hinchliffe (1981) and Hinchliffe and Gumpel-Pinot (1983).

In limb development, interdigital cell death shapes the emerging digits while the initial overproduction of the nerve supply to the limb is reduced at the spinal ganglia by the death of neurones that fail to contact muscle cells, intermediate neurones or sensory neurones (thus failing to obtain nerve growth factor). Cell death is a prominent feature of bird limb development (Saunders, Gasseling and Saunders, 1962) and the chick limb provides an ideal model. Three areas of cell death have been described in the mesenchyme of the early chick limb bud: the anterior and posterior necrotic zones (ANZ and PNZ) and a centrally located opaque patch (OP). Some cell death also occurs in the apical ectodermal ridge (AER), a thickening of ectoderm at the distal end of the limb bud (Fig. 6.2).

One suggestion made about the ANZ and PNZ is that they are involved in the reduction of the amount of tissue available for digit formation, since during evolution birds have undergone a process of

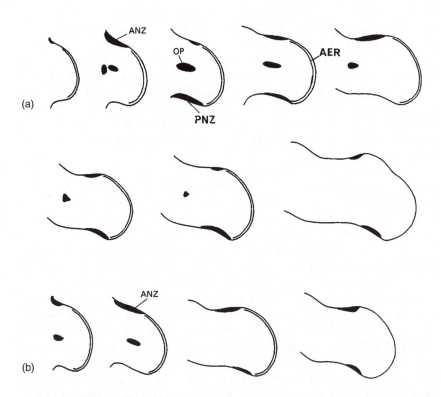

Fig. 6.2   The apical ectodermal ridge (AER) and the areas of mesenchymal cell death, the anterior necrotic zone (ANZ), posterior necrotic zone (PNZ) and opaque patch (OP) during the development of (a) a chick wing bud and (b) a herring gull wing bud. Source: redrawn from Hinchliffe (1981).

digital reduction from pentadactyl ancestors (they have only three wing digits, for example). The AER is considered to induce mesodermal outgrowth while in turn being dependent on a mesodermal maintenance factor. It is now thought that the overall anterio-posterior pattern of development is subject to a local hormone or morphogen, perhaps similar to retinoic acid, diffusing from a specific zone of mesoderm located along the posterior border of the limb bud, called the zone of polarizing activity (ZPA). If the ZPA is experimentally implanted in the pre-axial limb bud border it provokes formation of a second wing, whose axis it controls. Barrier and amputation experiments implicate the ZPA in the control of normal wing development, and exclusion of anterior parts of the wing bud from ZPA leads to the death of many of the distal cells underlying the AER. Other studies show that removal of the AER is followed by death of the underlying cells. Mutants confirm that the pattern of cell death obtained during morphogenesis is probably under genetic control. Thus, *wingless* mutants (*ws* gene) demonstrate an extension and expansion of cell death in the ANZ and lack a ZPA (Hinchliffe and Ede, 1973). Conversely, in *talpid* mutants there is an excessive elongation of the AER, a reduction in cell death and the formation of deformed polydactylous limbs.

An extensive pattern of programmed cell death occurs later on during limb development with the establishment of the so-called interdigital 'necrotic' zones at day 8 in the chick (Fig. 6.3). The type of cell death seen here is apoptotic not necrotic, although Hinchliffe has observed enhanced levels of autophagy in the fragmenting cells similar to that described in Clarke type 2 cell death. Interdigital cell death has been found in all amniote species, from mice to men. The zones may be absent or reduced in webbed species or *talpid* mutants. Interdigital cell death therefore plays an important role in moulding and sculpting the emergent pentadactyl limb. In this context, web-footed birds and mutants provide good controls for demonstrating the genetic basis of morphogenetic programmed cell death.

Parallel to the development of the limb bud, regulation of the neuronal cell number occurs in the relevant spinal ganglia. Neuronal cell number appears to be regulated by the degree of peripheral loading achieved (Prestige, 1970). Levi-Montalcini and Aloe (1981), reviewing the situation, showed that nerve growth factor (NGF) channelled via peripheral innervation played a significant role in controlling the number of cells in developing neuronal ganglia (see also Chapter 1). Experimental work on amphibians, chicks and mammals has confirmed the role of peripheral innervation and NGF feedback in controlling the level of cell death and thus cell survival. It has been shown that withdrawal of NGF induces programmed cell death in neurones, although this appears not to be strictly apoptotic (Martin and Johnson, 1991), the morphological profiles induced again being more akin to Clarke type 2 cell death.

Whatever the specific type of cell death induced by withdrawal of NGF, it is clearly genetically programmed, since the death can be prevented by inhibitors of RNA and protein synthesis (Fig. 6.3).

6.3.2 Cell death and differentiation in the vertebrate reproductive system

During the sexual differentiation of vertebrates, the reproductive system goes through a stage when both male and female embryos have the same structures. Later on in the male, the Wolffian duct differentiates into epididymis and vas deferens while the Müllerian duct regresses. In the female, the Müllerian duct differentiates into the uterus in mammals and the oviduct and shell gland in birds (Fig. 6.4).

Müllerian duct regression in the male provides an interesting example of the hormonal control of cell death. The type of cell death induced in the Müllerian ducts of both mouse and chick appears to be apoptotic, but

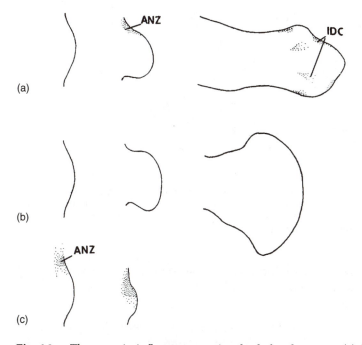

Fig. 6.3   The genetic influence on wing bud development. (a) Normal wing with cell death at the anterior necrotic zone (ANZ) and later interdigital cell death between the fingers. (b) *Talpid* mutant wing bud, where inhibition of cell death leads to excessive distal development. (c) *Wingless* mutant wing bud with enlarged ANZ, excessive cell death and little or no wing development. Source: redrawn from Hinchliffe, 1981.

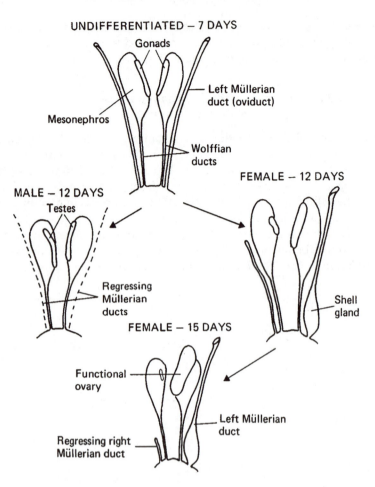

Fig. 6.4 Sexual differentiation in the chick embryo. The Müllerian duct (oviduct) regresses by means of hormonally induced cell death in male embryos, but survives and differentiates on the left side in female embryos. Source: redrawn from Hinchliffe (1981).

cell fragmentation does appear to be preceded by an enhancement of autophagic acid phosphatase and other lysosomal hydrolases.

6.3.3 Epithelial cell death during fusion of the secondary palate in mammalian development

A secondary palate separating the oral and nasal cavities develops by growth, rotation and fusion of left and right palatal shelves in mammals. This palatal fusion follows an adhesion between apposed epithelial surfaces, followed by the programmed death of cells in the midline, enabling the two shelves to fuse and developmentally complete a secondary palate.

6.4 CELL DEATH IN METAMORPHOSIS

Metamorphosis is a term employed to describe the major changes in shape, form and structure that occur during animal development. It is often used in the context of larval-to-adult transformations and as such is well represented in insects and amphibians, where large masses of tissue are required to degenerate rapidly and predictably, usually on a hormonal cue. The term is also used in the context of rapid embryological transformations in a range of examples where changes in overall structure, shape and form are involved. Such changes involve large-scale cell death and are seen in the development of most metazoa, but prove to be particularly striking in molluscs, echinoderms and ascidians. The developmental role of cell death in a range of such transformations has been extensively and ably reviewed by Glucksmann (1951) and Saunders (1966).

Perhaps the best examples of significant metamorphic cell death are seen in endopterygote insects and indeed, it is not surprising that where understanding of the control of such transformations has been combined with an extensive genetic database, as in *Drosophila*, breakthroughs have been made in uncovering the molecular and genetic basis of cell death. Since the genetic basis of programmed cell death in insects was covered in Chapter 4, attention here will be given to its incidence in amphibian metamorphosis and in particular to the disappearance of the tadpole tail. Interestingly, tadpole tail resorption during spontaneous metamorphosis was the subject of an early study of apoptosis by Kerr, Harmon and Searle (1974).

6.4.1 Tadpole tail metamorphosis

Tadpole tail resorption is initiated by rising levels of thyroxine. The cells of the tadpole respond differently to the rise in thyroxine such that only the cells in the tail degenerate. This suggests that the same cell types, in a tissue sense, have a different competence to respond in spatially different areas of the tadpole and supports the idea of a mosaic distribution of some gene-regulatory product earlier on in development (section 4.1). The first obvious sign following thyroxine climax is the collapse of the ventral and dorsal tail fins. The fins are relatively simple structures consisting of epidermis, melanocytes, fibroblasts, macrophages and hydrated collagen. One early effect of thyroxine is the activation of procollagenase and cathepsins in the tail, leading to a proteolytic cascade that destroys the tail fins. Thyroxine, in a complex way, appears to stimulate protein synthesis for degradation, leading to the *de novo* synthesis of hydrolytic enzymes. The programmed nature of the precipitated cell death is underlined by the fact that actinomycin D, an inhibitor of RNA synthesis, can stop the breakdown of experimentally isolated tadpole tail tip.

Tail fin collapse is followed by degradation of tail muscle, a catastrophic event recorded by many classical histologists and more recently by Weber (1969) and Fox (1973). The very extensive tail muscles fragment into smaller 'sarcolytes', which are subsequently engulfed and digested by macrophages in a process akin to apoptosis (Kerr, Harmon and Searle, 1974).

6.5 CELL DEATH IN PLANTS

An informed layman can see that in many respects plants are exercises in physiological cell death. Walking along an avenue of trees in autumn surrounded by so many leaves, so much seasonally programmed cell death, brings home the aptness of the term 'apoptosis', 'a falling away'. This massive orchestration of altruistic cell death may clearly serve to save the life of the tree in winter. Not only the nature of leaf fall, but the very transient nature of flowers, fruit and fruit ripening, a process often referred to as senescence, serve to emphasize the importance of programmed cell death in the botanical world. Even more significantly, wood, probably one of the first materials available to man, a source of shelter, warmth, defence and, indirectly, food, is the massive product of dead cells and the differentiation of xylem. Yet despite all this long-standing awareness and acceptance of the importance of cell death in plants, the field still inexplicably lags behind the burgeoning research in animal cell death. This may be partly due to the fact that plants usually produce rigid cell walls so that 'the cell' does not appear to go away when it dies and is often not disposed of. The dead nature of heartwood does not make it in any way less essential to the life of a tree than the dead cells that populate the cornified layers of the human skin.

Some progress has been made. The programmed nature of cell death in plants has been reviewed and scientifically explored (Gahan, 1981; Grierson, 1984; Woolhouse, 1984; Thomas, 1994; Schindler, Bergfield and Schopfer, 1995; Mittler, Shulaev and Lam, 1995). The work covered falls into three general areas of interest. The first and most basic deal with genes active in the cell cycle, the second with differentiation of xylem and phloem, and the third with senescence and cell-death genes.

6.5.1 Cell cycle and ageing genes

Most of this work deals with studies in yeasts, especially *Saccharomyces cerevisiae*. Here it has been shown that a particular set of cyclins, the G_1

cyclins, are limiting for passage through G_1 (Chapter 3). The *CLN3* gene product appears to initiate events that culminate in entry into the 'start' commitment point (Chasan, 1995). The activity of CLN3-CDC28 kinase is controlled by factors such as cell size. In animals the D-type cyclins seem to be involved in coupling the cell cycle to environmental signals such as growth factors (section 3.4). Jazwinski (1992), on the other hand, has identified a range of genes in yeast that control ageing and senescence, some of which may represent proto-suicide genes in an evolutionary sense. Mutation in *LAG1*, a longevity assurance gene, can shorten the replicative life span of yeast by 40%. Jazwinski has also shown that overexpression of *RAS2* extends the mean and maximum life span of yeast by 30%. It also appears to postpone ageing, as suggested by a delay in generation time. In yeast, RAS seems to integrate growth and cell division. It can sense nutritional status and activate either adenyl cyclase or inositol phospholipid turnover to stimulate cell growth and division. In contrast, under different conditions, it can pull them back by causing an arrest in the cell cycle. Senescing yeast arrest at the G_1/S boundary like human fibroblasts, where CDC7 plays a role in determining longevity.

Yeast is not a multicellular organism and as such does not enter into cell suicide. Primitive genes, influencing longevity and senescence in terms of their effects on the yeast cell cycle, may ultimately, however, prove to have some homology with genes that influence cell death in the metazoa and in this context they merit further study.

6.5.2 Differentiation and senescence to death in plants

In this context, the term **senescence** is used for the processes that lead ultimately to the death of an organism, organ or constituent cells. In a botanical context, authors speak of a seasonal senescence of the leaves alluded to earlier and clearly also cell senescence leading to cell death is involved in the formation of thorns, bark, the differentiation of phloem, xylem and heartwood, and the turnover of root hairs and root cap cells (Woolhouse, 1984).

Differentiation occurs as part of the senescence process and differentiation to death is common in plant tissues. Xylem vessels die in the course of their differentiation, in order to form tubular structures suitable for the transport of water. Xylem tissue itself is complex; it comprises parenchymatous cells, tracheids and fibres. The parenchymatous cells are long-lived and xylem fibres may also retain living protoplast after differentiation. Tracheids, on the other hand, invariably die in the course of their differentiation into xylem vessels.

It is still not entirely clear what controls xylem differentiation and the fate of particular cells, although it is thought that plant hormones or auxins are involved. It is also known that some bioregulators such as 6-benzyl amino purine (BAP) and 2,4-dichlorophenoxyacetic acid can induce xylogenesis.

During the process of xylogenesis, cells usually increase in size and this is followed by a process of lignification or thickening of the cellulose cell wall. During the process there is a progressive autolysis of cell contents and also a selective dissolution of particular areas of the cell wall.

The ordered arrangement of vascular tissue suggests that the differentiation involved is under genetic control, although direct evidence for this is not yet available. The fine-structural changes that occur in xylogenesis have been described and these include cellular enlargement and changes in nuclear shape and localized shrinkage of the cell walls, accompanied by hydrolytic activity in regions exposed to protoplast. These changes are followed by tonoplast (central vacuole) breakdown, dilation of the endoplasmic reticulum and detachment of ribosomes, finally accompanied by loss of mitochondria. The cytoplasm becomes highly vesicular. This morphological disruption is induced by a range of hydrolytic enzymes, including proteases.

The morphological changes, on the face of it, do not follow the sequence of events classically associated with apoptosis, although certain elements may be comparable. Gahan, Bowen and Winters (1995) have argued that some of the nuclear changes such as invagination and chromatin margination followed by cytoplasmic vacuolation make the processes broadly comparable. They also present preliminary evidence of endonuclease activity associated with chromatin margination and DNA degradation in the nucleus. Apoptotic-like behaviour was also reported by Mittler, Shulaev and Lam (1995) in nuclei from *Nicotiana tabacum*, and indeed, Gavrieli, Sherman and Ben-Sasson (1992) demonstrated *in situ* end-labelling of degraded DNA in developing tracheids and in induced lesions.

Changes similar to apoptosis have been reported in the nuclei of protophloem sieve-forming elements (Eleftheriou, 1986). Phloem, like xylem, is a complex tissue but mature sieve tubes are regarded as more or less senescent, in that during maturation the nuclei and some constituents of the cytoplasm are lost.

6.5.3 Aerenchyma formation

One form of cell death in plants that merits further investigation, if only for the reason that constitutive mutants are available, is aerenchyma formation in roots. Significant cell degeneration and death occurs in the roots of waterlogged plants. There is strong evidence that ethylene and hormonal induction is involved (Jackson, 1990). It appears that cortical cell lysis induces internal pathways that help to aerate roots as they extend into stagnant surroundings that are low in oxygen. It is thought that ethylene signals the very selective collapse of specific files of cells necessary to generate these pathways. Inhibitors of ethylene production or action inhibit aerenchyma formation when applied to poorly aerated roots.

During aerenchyma formation, cell breakdown begins with rupture of the tonoplast in cells some 10 mm behind the root apex and beyond the primodia or growing zone. This is followed by cell-wall degeneration. The fine structural basis of aerenchyma formation is similar in essence to that of xylogenesis, although it excludes lignification (Campbell and Drew, 1983). Little is known about the biochemistry, or indeed the control of these changes. Notably, however, cell death does not occur opposite lateral root primodia, suggesting that these particular cells are shielded from the cell-death pathway by the root primodia.

Aerenchyma formation should be amenable to genetic study, since in the roots of some cultivars of rice it appears to be a constitutive feature, not induced by waterlogging and ethylene. It would therefore be instructive to select mutants from among rice varieties that constitutively express this phenomenon. The pattern of cell death produced may not be entirely based on a genetically 'programmed senescence', since Elliot, Robson and Abbot (1993) point out that levels of phosphate and nitrogen affect the extent of cell death in root cortex.

6.5.4 The genetic basis of senescence and cell death in plants

Grierson (1984) concludes that studies on the synthesis of macromolecules in senescing tomatoes have provided evidence for the existence of a specific set of genes operating during ripening. One of these genes codes for polygalacturonidase, which plays a part in fruit softening and cell-wall degeneration. Indeed, it is now known that a wide range of genes operate in concert to progress the coordinated changes seen in plant senescence. Ethylene gas again seems to be implicated in the induction and expression of ripening genes.

Thomas (1994), reviewing ageing in plant and animal kingdoms, stresses the parallels that exist between programmed cell death in animal tissues and senescence and programmed cell death in plants. Thomas indicates that, while senescing plant cells are not dead, they are undergoing a phase of development that ends in death and this differentiation to death requires the coordinated orchestration of a range of genes showing specific parallels between senescence genes and genes that operate during programmed cell death in animals (Fig. 6.5).

Thomas (1987) has in fact identified a spontaneous mutation in a species of grass, *Festuca pratensis*, which has the effect of stabilizing the pigment–proteolipid complexes of thylakoid chloroplast membranes so that leaf tissue does not turn yellow during senescence. The *Sid* gene or

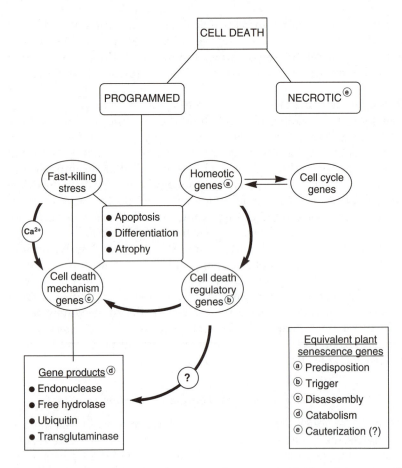

Fig. 6.5 Parallels between the genes that operate during programmed cell death in animals and cell senescence genes in plants. Source: redrawn from Thomas (1994)

'stay-green gene' is a Mendelian locus controlling thylakoid membrane disassembly in senescing leaves, and Thomas, Ougham and Davies (1992) went on to describe consistent differences in gene expression between normal genotype and lines homozygous or heterozygous for the mutant gene *Sid*. He identified, by subtractive means, a small number of new mRNA and polypeptide species in senescing leaves.

Kamachi *et al.* (1991) reported changing levels of cytosolic glutamine synthetase, chloroplastic glutamine synthetase, polypeptides and corresponding mRNAs during natural senescence in rice leaves. There appeared to be a significant transfer of glutamine from senescing leaves to growing tissues. The genetic basis of leaf senescence has been reviewed by Smart (1994).

The induction of leaf cell death by phytopathological bacteria has been described by Sigee (1984) and lastly, potential cell-death genes have been identified by Greenberg *et al.* (1994) in *Arabidopsis thaliana*, playing a role in a pathological response.

6.6 CELL DEATH IN THE THYMUS

The thymus provided the model on which the definition of apoptosis became based (Wyllie, 1981). Such a definition was broadly morphological, revolving around the shrinkage and condensation of cell and nucleus, accompanied by characteristic margination of chromatin and endonuclease activation (section 4.1). In fact, thymic cells are quite specialized and it is in some ways unfortunate that the definitions of apoptosis and indeed programmed cell death were initially based on such restricted material. Such a model, for example, probably puts undue emphasis on disposal of the nucleus and DNA and certainly led to an ill-advised diagnostic base for DNA laddering, which, as it happens, is not always characteristic of apoptosis and certainly not of programmed cell death in general.

In other respects, the choice of thymus was felicitous in that the thymus experiences a massive involution and cell death during development, apoptotic cell death being specifically triggered by glucocorticoid hormone at puberty.

There is some evidence to suggest that glucocorticoid sensitivity is genetically mediated. Ucker (1987) showed that cytotoxic T lymphocytes induce a cytolytic process in target cells, which, like glucocorticoid- or dexamethasone-mediated cytolysis of immature thymocytes, effects a rapid apoptotic-like cell death involving DNA fragmentation (Chapter 5). Using a thymoma mutant resistant to glucocorticoid- and CTL-induced cytolysis, he showed that reversion at a single gene locus restored sensitivity to both. This important finding, that thymic apoptosis may have a genetic base, was further supported by Duke, Sellins and Cohen (1988), who demonstrated that target-cell DNA fragmentation associated with apoptosis was not detected in nuclei incubated with CTL-derived lytic granules or CTL-derived endonuclease, leading them to conclude that CTL cells destroyed their target by triggering them to endogenously commit suicide.

It has also been shown that glucocorticoid-induced thymic apoptosis requires RNA and protein synthesis (Cohen and Duke, 1984). The apoptosis can be inhibited by cycloheximide and actinomycin D, which are protein and RNA synthesis inhibitors respectively. Endonuclease activation during thymic apoptosis requires both calcium and magnesium ions. Owens, Hahn and Cohen (1991) have shown that both radiation- and glucocorticoid-induced apoptosis require RNA and protein synthesis and that two significant 'death-associated' mRNAs are produced, RP-2 and RP-8. The latter, which is produced within 1 h of induction, has been shown to have a zinc finger domain suggestive of a DNA regulatory role and could thus be involved in switching on genes leading to a cascade of synthetic events (Chapter 1). The new specific mRNA molecules are not detectable after 6 h and are thus relatively short-lived; indeed, evidence is accumulating of a rapid degradation of ribosomal and messenger RNA in apoptotic thymic cells (Delic, Coppey-Moisan and Magdelenat, 1993).

It has been reported that thymocyte apoptosis can also be induced by calcium ionophores (e.g. A23187), phorbol esters (e.g. phorbol myristate acetate) and anti-CD3 antibody. Heat shock also appears to induce apoptosis in mouse thymocytes and protects them from glucocorticoid-induced cell death. More recently, biochemical studies have suggested that two enzymes normally involved in DNA replication and repair, topoisomerase 2 and poly-(ADP-ribose)-synthetase, may also play important roles in the mechanisms of programmed cell death (Chapter 5). Certainly, topoisomerase inhibitors have been shown to induce apoptosis in thymocytes (Onishi *et al.*, 1993). Clarke *et al.* (1993) showed that thymocyte apoptosis could be induced by p53-dependent and -independent pathways. They demonstrated that wild-type mouse thymocytes readily underwent apoptosis after treatment with ionizing radiation, glucocorticoid or etoposide, an inhibitor of topoisomerase 2, or after calcium-dependent activation of phorbol esters and calcium ionophore. In contrast, null p53 (p53-free) thymocytes were resistant to radiation and etoposide induction of apoptosis but sensitive to glucocorticoid and calcium induction. The results demonstrated that p53 exerts a significant and dose-dependent effect on the initiation of apoptosis in thymocytes, but only when it is induced by agents that cause DNA strand breaks (section 4.7.6).

One major function of the thymus and thymocytes is to establish an appropriate level of immune and autoimmune response.

6.6.1 Apoptosis and the autoimmune response

Apoptosis appears essential for the normal development, function and regulation of the immune system. The work of Smith *et al.* (1989) claims a role for the thymus and apoptosis in establishing an appropriate auto-immune response. Any cells carrying self-reactive receptors have to be eliminated in order to avoid autoimmunity. Thus, the immune system must recognize a variety of foreign antigens, but must be made to delete self-reactive effector cells (derived as T lymphocytes) that would otherwise cause autoimmune disease. This deletion takes the form of apoptosis in the thymus where up to 97% of thymocytes are deleted during the first few days of their life. In this complex weeding-out process, immature thymocytes are removed by apoptosis if they have an excess of self-recognizing receptors and also if they have no recognition receptors for the self MHC (major histocompatability complex for self-recognition), leaving a residue of immunologically appropriate cells.

Further apoptotic deletion of thymocyte-derived T lymphocytes occurs outside the thymus and an important mechanism here involves the activation of surface Fas molecules. Fas (CD95/Apo-1 receptor) is a member of the TNF receptor family and can transduce signals for either apoptosis or mitosis (Chapter 5). These are important elements in common with cytokine-mediated cell death in tumours (section 5.8.8).

Evidence is also emerging that apoptosis is involved in the mechanisms of B-lymphocyte selection.

6.6.2 Apoptosis and autoimmune disease

Autoimmune genes in mice have been shown to lead to defects in apoptosis. In humans, defects in the regulation of Fas/Apo-1 have also been associated with abnormal apoptosis. Mountz *et al.* (1994) have demonstrated clearly that autoimmune disease is partly a problem of defective apoptosis. Patients with autoimmune disease such as systemic lupus erythematosus, rheumatoid arthritis and scleroderma share in common an imbalance between the production and destruction of lymphocytes, synovial cells and fibroblasts respectively. Patients with systemic lupus erythematosus have increased levels of Fas that inhibit normal apoptosis, and mutations in genes affecting Fas, Fas ligand and haemopoietic cell phosphatase have been identified in diseased animal models. The oncogenes *bcl-2*, *p53* and c-*myc*, which help regulate apoptosis, are also abnormally expressed. Interestingly, potent inducers of apoptosis, including steroids, azathioprine, cyclophosphamide and methoxytrexate, are among the most effective therapies for autoimmune disease. Therapies that could specifically induce apoptosis in particular cellular populations without incurring side effects should improve future prospects for successful treatment of autoimmune disease.

Cells that are involved in the immune response include monocytes, neutrophils and eosinophils and these cells normally die within a relatively short period through apoptosis. This apoptosis, however, can be delayed or accelerated by a number of cytokines, including γ-interferon and TNF, leaving open the *in vivo* potential for prolonged inflammatory disease.

6.7 APOPTOSIS AND THE HAEMOPOIETIC SYSTEM

Cowling and Dexter (1994) have indicated that haemopoietic stem cells (blood-forming cells) can undergo apoptosis as a consequence of growth factor withdrawal.

In red blood cells, the hormone erythropoietin (EPO) induces differentiation in the form of haemoglobin synthesis in competent erythroid cells, which then enter a mitotic maturation phase. EPO also enhances mitotic proliferation of erythroid cells. It has now been shown that withdrawal of this growth factor leads to apoptosis.

In the case of white blood cells, the progenitor cells and mature cells such as neutrophils and eosinophils are sensitive to a growth factor called colony-stimulating factor (CSF). Removal of growth factor at any stage in haemopoietic development leads to the cessation of cell population growth and eventually, after a delay of about 20–30 h, the apoptotic death of the clone. Withdrawal of the cytokine IL-3 also induces apoptosis in haemopoietic cell lines. The mechanisms are not fully understood, but it is thought that the cytokine normally activates a central kinase, protein kinase C (PKC), which normally suppresses apoptosis in target cells. The associated cytoplasmic signalling proteins are not known but the inhibition of c-Fes, a member of the cytoplasmic tyrosine kinase family, does appear to enhance apoptosis (Manfredini *et al.*, 1973). In a paper emphasizing the multiple pathways to apoptosis, Evans (1993) stresses the probable central role of PKC in directing whether cells eventually enter mitosis or apoptosis in response to growth factor levels. Cowling and Dexter (1994) also indicate that enforced expression of *bcl-2* in haemopoietic stem cells inhibits apoptosis following growth factor withdrawal.

6.7.1 Apoptosis and mitosis in haematological disease

Changes in the balance between mitosis and apoptosis could have a clear bearing on haematological diseases such as anaemias and myeloid leukemia. Because of chromosomal translocations, leukaemic patients express a chimeric Bcr—Abl product, which elevates tyrosine kinase activity. Inhibition of tyrosine kinase activity leads to programmed cell death. It has been shown that expression of the gene *abl* is associated with a reduced level of apoptosis and an elevated level of mitosis at low growth factor levels (Spooner *et al.*, 1994).

6.8 MITOSIS AND APOPTOSIS IN THE GASTROINTESTINAL TRACT

Potten (1992) has shown that the gastrointestinal tract is highly structured and polarized, with a relatively few stem cells demonstrating a high level of mitosis. The stem cells are located in specific positions: at the crypts in the colon and about four cell positions from the base, above the Paneth cells in the intestine. He has also shown that a small but constant level of apoptosis occurs in the crypt. The levels of apoptosis are elevated by exposure to small doses of radiation and cytotoxic drugs. The maximum level of apoptosis seems to occur about 3–6 h after exposure and the cell death is characteristically located at the fourth cell level in the intestine. Anilkumar *et al.* (1992) report on the apoptosis induced in the intestinal crypts by a range of cancer chemotherapeutic agents, including cytosine arabinoside, vincristine and Adriamycin. They found that the compounds induced apoptosis in the proliferative compartment of the crypt and concluded that apoptosis can be induced by severe pathological perturbation as well as by more physiological triggers.

The intestinal villus therefore represents a complex but important model for the morphogenetic and homeostatic relationship between mitosis and apoptosis, the phenomena controlling cell proliferation and cell death respectively, and Potten (1992) goes on to speculate on the mechanisms underlying this control.

B. CELL DEATH IN TUMOURS

6.9 INTRODUCTION

Numerous examples of apoptosis in tumours have been dealt with in previous chapters. In Chapters 3 and 4, the role of oncogenes in mitosis and apoptosis was covered and genes affecting tumour growth and progression were referred to, especially *p53* and c-*myc*; the importance of Bcl-2 as an inhibitor of apoptosis in tumour cells was also mentioned. Chapter 5 dealt in detail with the role of cell killing and cytokines in precipitating apoptosis, where most of the target cells were tumour cell lines grown in tissue culture. In nature, however, tumours grow as a more or less solid ball of cells and have a very complex structure and development. Section A of this chapter dealt briefly with the tissue kinetics of tumours. This part will focus on the incidence and distribution of cell death, particularly apoptosis, in developing tumours and will go on to consider some therapeutic implications.

6.10 TUMOUR STRUCTURE

It is not the intention here to present an exhaustive overview of tumour structure. An excellent review of such subject matter has been published by Alison and Sarraf (1997). In a simplistic organizational sense, tumours may either be solid or corded (Fig. 6.6).

In the case of corded tumours, the younger, proliferative tumour cells subtend on a central blood vessel, much as liver parenchyma is arranged around central blood vessels. With respect to each cord, therefore, the older cells are pushed to the circumference or periphery and the younger tumour cells form a proliferative zone around the central blood vessels.

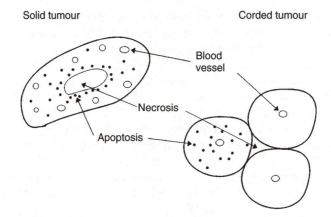

Fig. 6.6 The distribution of necrosis and apoptosis in solid and corded tumours. Note that in corded tumours the blood vessels are centrally placed with respect to the tumour cells, while the preponderance of blood vessels is peripheral in solid tumours. Necrosis occurs in areas of poor blood supply and is thus predominantly seen centrally in solid tumours and peripherally in corded tumours. In general, higher levels of apoptosis are also seen in ischaemically stressed areas of the tumour near to the necrosis.

In a solid tumour, the whole tumour takes up the form of a growing ball of cells and the younger proliferative population of tumour cells expands outwards, leaving the older and eventually necrotic cells in a more or less central position. In such a tumour, the younger tumour cells form the outer zone and engender their own vascular blood supply as they move outwards. Proliferating tumour cells produce a chemical called **angiogenesis factor** which induces the endothelial cells lining the blood capillaries to divide and branch, thus forming a new capillary bed, supplying the tumour cells with sufficient glucose and oxygen. The vascularization thus follows the tumour cells as they expand outwards. The older tumour cells produce less angiogenesis factor and the old capillary bed in the centre of the tumour is more easily compromised (Denekamp, Hill and Hobson, 1982). Because blood flow is poor in the centre of such a growing tumour, cells are often starved of glucose and oxygen and persistent anoxia will induce necrotic cell death (Chapter 1). Agents that can selectively compromise or interrupt the vascular blood supply to a tumour could thus have great therapeutic value.

It should be noted, however, that varying levels of cell deletion occur in the proliferative zones of both solid and corded tumours. The potential sources of these deaths are many, although close scrutiny indicates that they are largely due to apoptosis (Sarraf and Bowen, 1986, 1988) and occur naturally in untreated tumours. Interestingly, the incidence of apoptosis always appears somewhat higher in zones next to the necrotic areas, i.e. in somewhat hypoxic areas. In terms of causes, such single cell deletions could be due to a withdrawal of tumour cells to G_0 followed by subsequent differentiation and apoptotic death or could be the result of individual cell targeting by hunter–killer cells, including macrophages, NK cells, including the LAKs, CTLs and specific TILs. All these killer cells appear able in various ways (detailed in Chapter 5) to induce a target-mediated apoptosis in tumour cells. These methods include occupying Fas/Apo-1 receptors and also, in the case of CTL and LAK cells, the secretion of granzyme proteases, all leading to the activation of apoptosis in the targeted tumour cell. It is not surprising, therefore, that levels of naturally occurring apoptosis range between 1% and 10% in a wide spectrum of tumours and that from the therapeutic point of view the phenomenon deserves serious and urgent attention with a view to experimentally enhancing the levels of apoptosis.

6.11 TNFα AND TUMOUR CELL DEATH

One potentially therapeutic agent is tumour necrosis factor-alpha (TNFα), as the name suggests. However, as well as its inflammatory impact on tumour vascularization, TNF also induces apoptosis in sensitive tumour cells, as described in Chapter 5. Before returning to apoptosis, the impact of TNFα on haemorrhagic necrosis will be considered. It has been shown that TNFα is largely produced by macrophages in response to bacterial, viral or parasitological infection of the body. It also initiates extensive vascular changes in tumours. One major effect is to make the endothelial cells of the blood capillaries more adhesive, thus promoting platelet adhesion. This leads to blood clotting in the capillaries, blood stasis, lack of circulation and haemorrhage. In turn, this leads to oxygen starvation and necrosis of the surrounding tumour cells. It has recently been shown that the compounds 5,6-flavone-8-acetic acid and 5,6-dimethylxanthenone induce a decrease in blood flow in tumours leading to massive tumour necrosis and regression due to their ability to induce TNFα. TNFα appears to work in this case by stimulating cross-linking and adhesion between blood vessel endothelial cells that are induced to produce adhesion molecules ICAM-1 and VCAM-1, and leucocyte adhesion molecule ELAM-1. The therapeutic vascular shutdown induced by TNFα was reviewed by Mannel *et al.* (1995).

TNFα can also induce apoptosis in solid tumours. It is clear from current molecular studies that the reception and transduction mechanisms in TNFα-induced apoptosis can be quite complex (Chapter 5). Two genes expressing two receptors, *TNFR1* and *TNFR2*, have now been cloned and it is clear that occupation of these receptors, and indeed the Fas receptor, can, after intermediate transduction via other death-domain-containing proteins, lead to the activation of proteases (now called caspases), leading to apoptosis.

6.9.3 Killer-cells and apoptosis

It is interesting at this stage to note that the cytotoxic lymphocytes and activated natural killer cells produce their own proteases, called granzymes, which also, in association with a cell-membrane-perforating molecule (perforin), induce apoptosis in targeted tumour cells (Smyth and Trapani, 1995; Chapter 5).

From a therapeutic and clinical point of view, TNFα, although having all the requirements for precipitating massive tumour cell death, has proved somewhat disappointing because of its lethal inflammatory side effects. In humans, systemic TNFα can lead to circulation problems, oedema, endotoxic shock and death. Anchored to the macrophage membrane or targeted individually on a cell-to-cell basis, the molecule still has considerable promise. From this point of view, the selective activation or, indeed, clonal expansion of a range of target-oriented killer cells obtained from the patient (e.g. NK, CTL, TIL) and capable of inducing selective apoptosis on reintroduction holds great promise for the future treatment of tumours.

6.12 APOPTOSIS AND POTENTIAL TUMOUR THERAPY

The absence of p53 or aberrant p53 means that apoptosis cannot be engaged even when the cell receives genetic damage (Chapter 4). Indeed, aberrant p53 seems to be an element in carcinogenesis, since faulty DNA cannot be eliminated through the normal p53-activated apoptotic pathway and the majority of solid tumours, particularly lung, colon and breast, are either missing p53 or produce a faulty version. Kindlier and Vogelstein (1996), describing life and death in a malignant tumour, emphasize the role of hypoxic *p53* mutants in the development of apoptotically resistant tumour and metastases. Many tumour cells also appear insensitive to growth factor withdrawal, so cannot enter a p53-independent apoptotic pathway; others are insensitive to immune and Fas-relayed apoptosis. It has therefore been argued that tumours are as much a result of apoptotic failure as of a deregulation of the cell cycle. Therapeutic approaches include the introduction of a normal *p53* gene into cancers as part of an integrated gene therapy and the blocking of growth factor reception or transduction.

Similar problems revolve around *bcl-2*, which is upregulated in certain lymphomas (Chapter 4). Excess Bcl-2 production will in itself inhibit apoptosis. Some vital cells, e.g. melanocytes, normally have elevated levels of Bcl-2, and here, if the cells become cancerous, they are less susceptible to apoptosis, producing an aggressive melanoma. In this instance, repression of the *bcl-2* gene or neutralization of the product, Bcl-2, could be therapeutic. This topic is related to the oncogenic viruses that inhibit apoptosis; thus, the Epstein–Barr virus BHRFI protein, a homologue of Bcl-2, inhibits apoptosis, and LMP-1 protein from the same virus induces a zinc finger protein that inhibits TNF cytotoxicity. The adenovirus *E1B* gene also encodes a protein homologue of Bcl-2 that inhibits p53. Viruses not only inhibit apoptosis, of course, but can also induce cell proliferation.

Adenoviruses are known to activate cells from growth arrest, probably through inactivation of Rb protein, release of transcription factor E2F and activation of c-*myc*.

One relatively new development relates to the finding that kinase inhibitors can produce tumour cell death. In a provocative paper, Szende *et al.* (1995) have demonstrated the cell-death-inducing effect of an EGF receptor tyrosine kinase inhibitor, alpha-cyano-3,4-dihydroxycinnanthiomide. Protein and RNA synthesis inhibitors block the programmed cell death, which morphologically resembles Clarke type 3 cell death rather than apoptosis. Similar results are reported by Szegedi *et al.* (1996) in colon carcinoma cells and also independently by Mason *et al.* (personal communication), using another kinase inhibitor affecting tumour cell adhesion. What is interesting about these results is that the cells vacuolate and swell, showing changes in hydrolase distribution as reported in invertebrate programmed cell death, rather than shrinking, as in apoptosis. Cell–matrix interactions are also apparently as important in inducing apoptosis as cell–cell interactions, and Frisch and Francis (1994) have shown that an absence of integrin-based matrix can induce apoptosis. Thus, homelessness can lead to cell suicide.

Finally, more drastic direct methods can be resorted to in order to induce cell death in tumours. Irradiation, chemotherapy and appropriate hormone therapy can all induce apoptosis, although the higher doses may of course kill cells through necrosis. The production of limited DNA damage induces apoptosis, mediated via p53. The role of apoptosis in cancer chemotherapy is usefully reviewed by Hickman *et al.* (1994) and the role of apoptosis in health and disease is exhaustively reviewed by Bellamy *et al.* (1995).

Lastly, it may be worth reflecting once more on the original paradox that the whole of metazoan life is dependent on death and noting that the life of every cancer patient can potentially be saved by enhancing cell death.

References

Abrams, J. M., White, K., Fessler, L. I. and Steller, H. (1993) Programmed cell death during Drosophila embryogenesis. *Development*, **117**, 29–43.

Aherne, W. A., Camplejohn, R. S. and Wright, N. A. (1977) *Cell Population Kinetics*, Edward Arnold, London.

Akbar, A. N., Borthwich, N., Salmon, M. *et al.* (1993) The significance of low *bcl-2* expression by CD45RO T-cells in normal individuals: the role of apoptosis in T cell memory. *J. Exp. Med.*, **178**, 427–438.

Akbar, A. N., Borthwich, N. J., Wickremaisinghe, R. G. *et al.* (1996) Interleukin-2 receptor common gamma-chain signaling cytokines regulate activated T-cell apoptosis in response to growth factor withdrawal: selective induction of anti-apoptotic (*bcl-2*, *bcl-X$_L$*) but not pro-apoptotic (*bax*, *bcl-X$_S$*) gene expression. *Eur. J. Immunol.*, **26**, 294–99.

Alison, M. and Sarraf, C. (1997) *Understanding Cancer from Basic Science to Clinical Practice*, Cambridge University Press, Cambridge.

Alnemri, E. S., Livingston, D. J., Nicholson, D. W. *et al.* (1996) Human ICE/CED3 protease nomenclature. *Cell*, **87**, 171.

Amati, B., Littlewood, T., Evan, G. and Land, H. (1994) The c-Myc protein induces cell cycle progression and apoptosis through dimerisation with Max. *EMBO J.*, **12**, 5083–5087.

Anilkumar, T. V., Sarraf, C. E., Hunt, T. and Alison, M. R. (1992) The nature of cytotoxic drug-induced cell death in murine intestinal crypts. *Br. J. Cancer*, **65**, 556–558.

Arcaro, A. and Wymann, M. P. (1993) Wortmannin is a potent phosphatidyl-inositol 3-kinase inhibitor: the role of phosphatidylinositol 3,4,5-trisphos-phate in neutrophil responses. *Biochem J.*, **296**, 297–301.

Arends, M. J. and Wyllie A. H. (1991) Apoptosis mechanisms and roles in pathol-ogy. *Int. Rev. Exp. Pathol.*, **32**, 223–254.

Arteaga, C. L., Hurd, S. D., Winnier A. R. *et al.* (1993) Anti-transforming growth factor (TGF)-β antibodies inhibit breast cancer cell tumorigenicity and increase mouse spleen natural killer cell activity: implications for a possible role of tumour cell/host TGF-β interactions in human breast cancer progression. *J. Clin. Invest.*, **92**, 2569–2576.

Askew, D. S., Ashman, R. A., Simmons, B. C. and Cleveland, J. L. (1991) Constitutive c-*myc* expression in an IL3 dependent myeloid cell line suppresses cell cycle arrest and accelerates apoptosis. *Oncogene*, **6**, 1915–1922.

Ayoub, I. A. and Yang, T. J. (1996) Growth regulatory effects of transforming growth factor-β1 and interleukin 2 in IL-2 dependent CD4$^+$ T lymphoblas-toid cell line. *Immunol. Invest.*, **25**, 129–151.

Baringa, M. (1996) Forging a path to cell death. *Science*, **273**, 735—737.

Bazan, J. F. (1993) Emerging families of cytokines and receptors. *Curr. Biol.*, **3**, 603–606.

Beidler, D., Tewari, M., Freesen, P. *et al.* (1995) The baculovirus p35 protein inhibits Fas and tumour necrosis factor induced apoptosis. *J. Biol. Chem.*, **270**, 16526–16528.

Bellamy, C. O. C., Malcomson, R. D. G., Harrison, D. J. and Wyllie, A. H. (1995) Cell death in health and disease: the biology and regulation of apoptosis. *Cancer Biol.*, **6**, 3–16.

Bellone, G., Aste-Amezaga, M., Trinchieri, G. and Rodeck, U. (1995) Regulation of NK cell function by TGF-β1. *J. Immunol.*, **155**, 1066—1073.

Berke, G. (1997) Killing mechanism of cytotoxic lymphocytes. *Curr. Opin. Hematol.*, **4**, 32–40.

Beutler, B. and Cerami, A. (1986) Cachectin and tumour necrosis factor as two sides of the same biological coin. *Nature*, **320**, 584–588.

Biose, L., Gonzalez-Garcia, M., Postema, C. *et al.* (1993) bcl-x, a bcl-2 related gene that functions as a dominant regulator of apoptotic cell death. *Cell*, **74**, 597–608.

Bissonnette, R. P., Echeverri, F., Mahboubi, A. and Green, D. R. (1992) Apoptotic cell death induced by c-*myc* is inhibited by bcl-2. *Nature*, **359**, 552–556.

Bourne, H. R., Sanders, D. A. and McCormick, F. (1991) The GTPase superfamily: conserved structure and molecular mechanism. *Nature*, **349**, 117–127.

Bowen, I. D. (1993) Apoptosis or programmed cell death? *Cell Biol. Int.*, **17**, 365–380.

Bowen, I. D. and Bowen S. M. (1990) *Programmed Cell Death in Tumours and Tissues*, Chapman & Hall, London and New York.

Bowen, I. D., Den Hollander, J. E. and Lewis, G. H. J. (1982) Cell death and acid phosphatase activity in the regenerating planarian *Polycelis tenuis*. *Differentiation*, **21**, 160–170.

Bowen, I. D. and Lockshin, R. A. (1981) *Cell Death in Biology and Pathology*, Chapman & Hall, London.

Bowen, I. D., Morgan S. M. and Mullarkey, K. (1993) Cell death in the salivary glands of metamorphosing *Calliphora vomitoria*. *Cell Biol. Int.*, **17**, 13–33.

Bowen, I. D., Mullarkey, K. and Morgan, S. M. (1996) Programmed cell death in the salivary glands of the blow fly *Calliphora vomitoria*. *Microsc. Res. Tech.*, **34**, 202–207.

Bowen, I. D. and Ryder, T. A. (1974) Cell autolysis and deletion in the planarian *Polycelis tenuis* Iijima. *Tissue Cell Res.*, **154**, 265–274.

Bowen, I. D. and Ryder, T. A. (1976) Use of the p-nitrophenyl phosphate method for the demonstration of acid phosphatase during starvation and cell autolysis in the planarian *Polycelis tenuis* Iijima. *Histochem. J.*, **8**, 319–329.

Bowen, I. D., Ryder, T. A. and Dark, C. (1976) The effects of starvation on the planarian worm *Polycelis tenuis*. *Cell Tissue Res.*, **169**, 193–206.

Boyd, J. M., Malstram, S., Subramanian, T. *et al.* (1994) Adenovirus, E1B 19kDa and bcl-2 proteins interact with a common set of cellular proteins. *Cell*, **79**, 341–351.

Broome, H.-E., Dargan, C. M., Krajewski, S. and Reed, J. C. (1995) Expression of Bcl-2; Bcl-X, and Bax after T-cell activation and IL-2 withdrawal. *J. Immunol.*, **155**, 2311–2317.

Brugarolas, J., Chandrasekavan, C., Gardon, J. I. *et al.* (1995) Radiation induced cell cycle arrest compromised by p21 deficiency. *Nature*, **377**, 552–557.

Brunda, M. J., Laistro, L., Warrier, R. R. *et al.* (1994) Antitumor and antimetastatic activity of interleukin-12 against murine tumors. *J. Exp. Med.*, **178**, 1223–1230.

Campbell, R. and Drew, M. C. (1983) Electron microscopy of gas space (aerenchyma) formation in the adventitious roots of *Zea mays* L. subjected to oxygen shortage. *Planta*, **157**, 350–375.

Canman, C. E., Gilmer, T. M., Coutts, S. B. and Kastan, M. B. (1995) Growth factor modulation of p53 mediated growth arrest versus apoptosis. *Genes Dev.*, **9**, 600–611.

Carswell, E. A., Old, L. J., Kassel, R. L. *et al.* (1975) An endotoxin-induced serum factor that causes necrosis of tumours. *Proc. Natl Acad. Sci. USA*, **72**, 3666–3670.

Casciola-Rosen, L., Nicholson, D. W., Chong, T. *et al.* (1996) Apopain/CPP32 cleaves proteins that are essential for cellular repair: a fundamental principle of apoptotic cell death. *J. Exp. Med.*, **183**, 1957–1964.

Chaouchi, N., Wallon, C., Goujard, C. *et al.* (1996) Interleukin-13 inhibits interleukin-2-induced proliferation and protects chronic lymphocytic leukemia B cells from *in vitro* apoptosis. *Blood*, **87**, 1022–1029.

Chasan, R. (1995) STARTing the plant cell cycle. *Plant Cell*, **7**, 1–4.

Chen, G., Shi, L., Litchfield, D. W. and Greenberg, A. H. (1995) Rescue from granzyme B-induced apoptosis by Wee 1 kinase. *J. Exp. Med.*, **181**, 2295–2300.

Cheng, E. H. Y., Levine, B., Boise, L. H. *et al.* (1996) Bax-independent inhibition of apoptosis by Bcl-X$_L$. *Nature*, **379**, 554–556.

Chinnaiyan, A. M., O'Rourke, K., Tewari, M. and Dixit, V. M. (1995) FADD, a novel death domain-containing protein, interacts with the death domain of Fas and initiates apoptosis. *Cell*, **81**, 505–512.

Chinnaiyan, A. M., Tepper, C. G., Seldin, M. F. *et al.* (1996) FADD/MORT1 is a common mediator of CD95 (Fas/APO-1) and tumor necrosis factor receptor-induced apoptosis. *J. Biol. Chem.*, **271**, 4961–4965.

Chinnaiyan, A. M., Chaudhary, D., O'Rourke, K. *et al.* (1997a) Role of CED-4 in the activation of CED-3. *Nature*, **388**, 728–729.

Chinnaiyan, A.M., O'Rourke, K., Lane, B.R. and Dixit, V.M. (1997b) Interaction of CED-4 with CED-3 and CED-9: a molecular framework for cell death. *Science*, **275**, 1122–1126.

Clarke, P. G. H. (1990) Developmental cell death: morphological diversity and multiple mechanisms. *Anat. Embryol.*, **181**, 195–206.

Clarke, A. R., Purdie, C. A., Harrison, D. J. *et al.* (1993) Thymocyte apoptosis induced by p53-dependent and independent pathways. *Nature*, **363**, 849–852.

Clem, R. J., Fechheimer, M. and Miller L. K. (1991) Prevention of apoptosis by a baculovirus gene during infection of insect cells. *Science*, **254**, 1388–1390.

Cohen, J. J. and Duke, R. C. (1984) Glucocorticoid activation of a calcium-dependent endonuclease in thymocyte nuclei leads to cell death. *J. Immunol.*, **132**, 38–42.

Collins, M. K. L., Marvel, J., Malde, P. and Lopez-Rivas, A. (1992) Interleukin 3 protects murine bone marrow cells from apoptosis induced by DNA damaging agents. *J. Exp. Med.*, **171**, 1043–1051.

Colombel, M., Olsson, C. A., Ng, P.-Y. and Buttyan, R. (1992) Hormone regulated apoptosis results from re-entry of differentiated prostate cells into a defective cell cycle. *Cancer Res.*, **52**, 4313–4319.

Cotter, T. G., Lennon, S. V. and Martin, S. J. (1990) Apoptosis: programmed cell death. *J. Biomed. Sci.*, **2**, 72–80.

Courvalin, J.-C., Segil, N., Blobel, G and Worman, H. (1992) The lamin B receptor of the inner nuclear membrane undergoes mitosis-specific phosphorylation and is a substrate for p34^{cdc2}-type protein kinase. *J. Biol. Chem.*, **67**, 19035–19038.

Cowling, G. J. and Dexter, T. M. (1994) Apoptosis in the haemopoietic system. *Phil. Trans. R. Soc. Lond. B*, **345**, 275–263.

Cuvillier, O., Pirianov, G., Kleuser, B. *et al.* (1996) Suppression of ceramide-mediated programmed cell death by sphingosine-1-phosphate. *Nature*, **381**, 800–803.

Darmon, A. J., Nicholson, D. W. and Bleackley, R. C. (1995) Activation of the apoptotic protease CPP32 by cytotoxic T-cell derived granzyme B. *Nature*, **377**, 446–448.

Delic, J., Coppey-Moisan, M. and Magdelenat, H. (1993) Gamma-ray-induced transcription and apoptosis-associated loss of 28S rRNA in interphase human lymphocytes. *Int. J. Rad. Biol.*, **64**, 39–46.

Denekamp, J., Hill, S. A. and Hobson, B. (1982) Vascular occlusion and tumour cell death. *Eur. J. Cancer Clin. Oncol.*, **19**, 271–275.

Deng, C., Zhang, P., Harper, J. W. *et al.* (1995) Mice lacking p21$^{pc11/waf1}$ undergo normal development, but are defective in G$_1$ checkpoint control. *Cell*, **82**, 675–684.

Di Virgilio, F., Pizzo, P., Zanovello, P. *et al.* (1990) Extracellular ATP as a possible mediator of cell-mediated cytotoxicity. *Immunol. Today*, **11**, 274–276.

Douglas, K. (1994) Making friends with death-wish genes. *N. Sci.*, **July**, 31–34.

Duan, H., Chinnaiyan, A. M., Hudson, P. L. *et al.* (1996a) ICE-LAP3, a novel mammalian homolog of the *Caenorhabditis elegans* cell death protein CED-3, is activated during Fas- and tumor necrosis factor-induced apoptosis. *J. Biol. Chem.*, **271**, 35013–35035.

Duan, H., Orth, K., Chinnaiyan, A. M. *et al.* (1996b) ICE-LAP6, a novel member of the ICE/*ced-3* gene family, is activated by the cytotoxic T-cell protease granzyme B. *J. Biol. Chem.*, **271**, 16720–16724.

Duke, R. C., Ojcius D. M. and Young, J. D. E. (1996) Cell suicide in health and disease. *Sci. Am.*, **Dec.**, 48–55.

Duke, R. C., Sellins, K. S. and Cohen, J. J. (1988) Cytolytic lymphocyte-derived lytic granules do not induce DNA fragmentation in target cells. *J. Immunol.*, **141**, 2191–2194.

Edgar, B. A. and O'Farrell, P. A. (1989) Genetic control of cell division patterns in the *Drosophila* embryo. *Cell*, **57**, 177–187.

Egan, S. E., Giddings, B. W., Brooks, M. W. *et al.* (1993) Association of Sos Ras exchange protein with Grb2 is implicated in tyrosine kinase signal transduction and transformation. *Nature*, **363**, 45–51.

Eissner, G., Kohlhuber, F., Grell, M. *et al.* (1995) Critical involvement of transmembrane tumour necrosis factor-α in endothelial programmed cell death mediated by ionizing radiation and bacterial endotoxin. *Blood*, **86**, 4184–4193.

Eleftheriou, E. P. (1986) Ultrastructural studies on protophloem sieve elements in *Triticum aestivum* L. nuclear degeneration. *J. Ultrastruct. Mol. Struct. Res.*, **95**, 47–54.

Elliot, G. A., Robson, A. D. and Abbot, L. K. (1993) Effects of phosphate and nitrogen application on death of the root cortex in spring wheat. *N. Phytol.*, **123**, 375–382.

Ellis, H. M. and Horovitz, H. R. (1986) Genetic control of programmed cell death in *C. elegans*. *Cell*, **44**, 817–829.

Ellis, R. E., Yuan, J. and Horovitz, H. R. (1991) Mechanisms and functions of cell death. *Annu. Rev. Cell. Biol.*, **7**, 663–698.

Enari, M., Hug, H. and Nagata, S. (1995) Involvement of an ICE-like protease in Fas-mediated apoptosis. *Nature*, **375**, 78–81.

Enari, M., Talanian, R. V., Wong, W. W. and Nagata, S. (1996) Sequential activation of ICE-like and CPP32 like proteases during Fas-mediated apoptosis. *Nature*, **380**, 723–726.

Evan, G. I. and Littlewood, T. D. (1993) The role of c-*myc* in cell growth. *Curr. Opin. Genet. Dev.*, **3**, 44–49.

Evan, G. I., Harrington, E., Fanidi, A. *et al.* (1994) Integrated control of cell proliferation and cell death by the c-*myc* oncogenes. *Phil. Trans. R. Soc. Lond. B*, **345**, 269–275.

Evan, G. I., Wyllie, A. H., Gilbert, C. S. *et al.* (1992) Induction of apoptosis in fibroblasts by c-myc protein. *Cell*, **69**, 119–128.

Evans, V. G. (1993) Multiple pathways to apoptosis. *Cell Biol. Int.*, **17**, 461–476.

Fagan, R., Flint, K. and Jones, N. (1994) Phosphorylation of E2F-1 modulates its interaction with the retinoblastoma gene product and the adenovirus E4 19kDa protein. *Cell*, **78**, 799–811.

Farrow, S. N., White, J. H. M., Martinou, I. *et al.* (1995) Cloning of a *bcl-2* homologue by interaction with adenovirus E1B 19K. *Nature*, **374**, 731–733.

Faucheu, C., Diu, A., Chan, A. W. E. *et al.* (1995) A novel human protease similar to the interleukin-1β converting enzyme induces apoptosis in transfected cells. *EMBO J.*, **14**, 1914–1922.

Fernandes-Alnemri, T., Litwack, G. and Alnemri, E. S. (1994) CPP-32, a novel human apoptotic protein with homology to *Caenorhabditis elegans* cell death protein CED-3 and mammalian interleukin-1β-converting enzyme. *J. Biol. Chem.*, **269**, 30761–30764.

Fernandes-Alnemri, T., Litwack, G. and Alnemri, E. S. (1995) Mch-2, a new member of the apoptotic CED-3/ICE cysteine protease gene family. *Cancer Res.*, **55**, 2737–2742.

Fernandes-Alnemri, T., Takahaski, A., Armstrong, R. *et al.* (1995) Mch-3, a novel human apoptotic cysteine protease highly related to CPP32. *Cancer Res.*, **55**, 6045–6052.

Fernandes-Alnemri, T., Armstrong, R. C., Krebs, J. *et al.* (1996) *In vitro* activation of CPP-32 and Mch3 by Mch4, a novel human apoptotic cysteine protease containing two FADD-like domains. *Proc. Natl Acad. Sci. USA*, **93**, 7464–7469.

Fisher, R. P. and Moran, D. O. (1994) A novel cyclin associates with MO15/cdk7 to form the cdk-activating kinase. *Cell*, **78**, 713–724.

Fluckiger, A.-C., Durand, I. and Bauchereau, J. (1994) Interleukin-10 induces apoptotic cell death of B-chronic lymphocytic leukemia cells. *J. Exp. Med.*, **179**, 91–99.

Fox, H. (1973) Ultrastructure of tail degeneration in *Rana temporaria* larva. *Fol. Morphol.*, **21**, 109–112.

Fraser, A. and Evan, G. (1996) A licence to kill. *Cell*, **85**, 781–785.

Frisch, S. M. and Francis, H. (1994) Disruption of epithelial cell–matrix interactions induces apoptosis. *J. Cell Biol.*, **124**, 619–626.

Fristrom, M. (1972) Chemical modification of cell death in the bar eye of *Drosophila*. *Mol. Gen. Genet.*, **115**, 10–18.

Froelich, C. J., Hanna, W. L., Poirier, G. G. *et al.* (1996) Granzyme B/perforin-mediated apoptosis of Jurkat cells results in damage of poly (ADP-ribose) polymerase to the 89-kDa apoptotic fragment and less abundant 64-kDa fragment. *Biochem. Biophys. Res. Commun.*, **227**, 658–665.

Furuya, Y., Berges, R. S., Lundmo, P. and Isaacs, J. T. (1994) Proliferation independent activation of programmed cell death as a novel therapy for prostrate cancer, in *Apoptosis* (eds E. Mihich and R. T. Schimke), Plenum Press, New York.

Gahan, P. B. (1981) Cell senescence and death in plants, in *Cell Death in Biology and Pathology* (eds I. D. Bowen and R. A. Lockshin), Chapman & Hall, London, pp. 145–169.

Gahan, P. B., Bowen, I. D. and Winters, C. (1995) Plant bioregulator-induced apoptotic-like behaviour of nuclei from mesophyll cells of *Salanium avicu-lare*. *Proc. R. Microsc. Soc.*, **30**, 117–118.

Gajewski, T. F. and Thompson, C. B. (1996) Apoptosis must signal transduction: elimination of a BAD influence. *Cell*, **87**, 589–592.

Gavrieli, Y., Sherman, Y. and Ben-Sasson, S. A. (1992) Identification of programmed cell death *in situ* via specific labelling of nuclear DNA fragmentation. *J. Cell Biol.*, **119**, 493–501.

Gearing, D. P. and Ziegler, S. F. (1993) The haematopoietic growth factor receptor family. *Curr. Opin. Hemat.*, **17**, 138–148.

Gehring, W. J. (1987) Homeo boxes in the study of development. *Science*, **236**, 1245–1252.

Gerard, C., Brayns, C., Marchant, A. *et al.* (1993) Interleukin-10 reduces the release of tumour necrosis factor and prevents lethality in experimental endotoxemia. *J. Exp. Med.*, **177**, 547–550.

Ghayur, T., Banerjee, S., Hugunin, M. *et al.* (1997) Caspase-1 processes IFN-γ-inducing factor and regulates LPS-induced IFN-γ production. *Nature*, **386**, 619–623.

Glucksmann, A. (1951) Cell deaths in normal vertebrate ontogeny. *Biol. Rev.*, **26**, 59–86.

Goodrich, D. and Lee, W. (1993) Molecular characterization of the retinoblastoma susceptibility gene. *Biochim. Biophys. Acta*, **1155**, 43–61.

Greenberg, J. T., Guo, A., Klessig, D. F. and Ausubel, M. (1994) Programmed cell death in plants: a pathogen-triggered response activated coordinately with multiple defence functions. *Cell*, **77**, 551–563.

Grierson, D. (1984) Nucleic acid and protein synthesis during fruit ripening and senescence, in *Cell Ageing and Cell Death* (eds I. Davies and D. C. Sigee), S. E. B. Seminar Series 25, Cambridge University Press, Cambridge, pp. 189–202.

Gu, Y., Sarnecki, C., Aldape, R. A. *et al.* (1995) Cleavage of poly (ADP-ribose) polymerase by interleukin-1β converting enzyme and its homologs TX and Nedd2. *J. Biol. Chem.*, **270**, 18715–18718.

Haldar, S., Jena, N. and Croce, C. M. (1995) Inactivation of Bcl-2 by phosphorylation. *Proc. Natl Acad. Sci. USA*, **92**, 4507–4511.

Hanada, M., Aimé-Sempé, C., Sato, T. and Reed, J. C. (1995) Structure–function analysis of bcl-2 protein: identification of conserved domains important for homodimerization with bcl-2 and heterodimerization with bax. *J. Biol. Chem.*, **270**, 11962–11968.

Harrington, E. A., Fanidi, A. and Evan, G. I. (1994) Oncogenes and cell death. *Curr. Opin. Genet. Dev.*, **4**, 120–129.

Hartwell, L., Culotti, J., Pringle, J. and Reid, B. (1974) Genetic control of the cell division in yeast. *Science*, **183**, 46–51.

Hay, B. A., Wasserman, D. A. and Rubin, G. M. (1995) Drosophila homologs of Baculovirus inhibitor of apoptosis proteins function to block cell death. *Cell*, **83**, 1253–1262.

Hengartner, M. O. (1997) CED-4 is a stranger no more. *Nature*, **388**, 714–715.

Hengartner, M. O., Ellis, R. E. and Horovitz H. R. (1992) *Caenorhabditis elegans* gene *ced-9* protects cells from programmed cell death. *Nature*, **356**, 494–499.

Hengartner, M. O. and Horovitz, H. R. (1994a) Activation of *C. elegans* cell death protein CED-9 by an amino acid substitution in a domain conserved in Bcl-2. *Nature*, **369**, 318–320.

Hengartner, M. O. and Horovitz, H. R. (1994b) *C. elegans* cell survival gene *ced-9* encodes a functional homolog of the mammalian proto-oncogene *bcl-2*. *Cell*, **76**, 665–676.

Hengartner, M. O. and Horovitz, H. R. (1994c) The ins and outs of programmed cell death during *C. elegans* development. *Phil. Trans. R. Soc. Lond. B*, **345**, 243–256.

Hickman, J. A., Potten, C. S., Merritt, A. J. and Fisher, T. C. (1994) Apoptosis and cancer chemotherapy. *Phil. Trans. R. Soc. Lond. B*, **345**, 319—325.

Hinchliffe, J. R. (1981) Cell death in embryogenesis, in *Cell Death in Biology and*

Pathology (eds I. D. Bowen and R. A. Lockshin), Chapman & Hall, London, pp. 35–78.

Hinchliffe, J. R. and Ede, D. A. (1973) Cell death and the development of limb form and skeletal pattern in normal and wingless (*ws*) chick embryos. *J. Embryol. Exp. Morphol.*, **30**, 753–772.

Hinchliffe, J. R. and Gumpel-Pinot, M. (1983) Experimental analysis of avian limb morphogenesis, in *Current Ornithology*, vol. 1 (ed. R. F. Johnston), Plenum Publishing, New York, pp. 293–327.

Hinds, P. W. and Weinberg, R. A. (1994) Tumor suppressor genes. *Curr. Opin. Genet. Dev.*, **4**, 135–141.

Horovitz, H. R. (1994) Genetic control of programmed cell death in the nematode *Caenorhabditis elegans*, in *Apoptosis* (eds E. Mihich and R. T. Schimke), Plenum Press, New York, pp. 1–11.

Hsu, H., Xiong, S and Goeddel, D. V. (1995) The TNF receptor 1-associated protein TRADD signals death and NF-κB activation. *Cell*, **81**, 495–504.

Hsu, H., Shu, H.-B., Pan, M.-P. and Goeddel, D. V. (1996) TRADD-TRAF2 and TRADD-FADD interactions define two distinct TNF receptor-1 signal transduction pathways. *Cell*, **84**, 299–308.

Hudig, D., Ewoldt, G. R. and Woodward, S. L. (1993) Proteases and lymphocyte cytotoxic killing mechanisms. *Curr. Opin. Immunol.*, **5**, 90–96.

Hunter, J., Bond, B. and Parslow, T. (1996) Functional dissection of the human Bcl-2 protein: sequence requirements for inhibitor of apoptosis. *Mol. Cell Biol.*, **16**, 877–883.

Igarashi, M., Nagata, A., Jinno, S. *et al.* (1991) Wee1[+]-like gene in human cells. *Nature*, **353**, 80–82.

Ingham, P. W., Howard, K. R. and Ish-Horowicz, D. (1985) Transcription pattern of the *Drosophila* segmentation gene hairy. *Nature*, **318**, 439–445.

Jäättelä, M., Benedict, M., Tewarl, M. *et al.* (1995) Bcl-x and Bcl-2 inhibit TNF and Fas-induced apoptosis and activation of phospholipase A_2 in breast carcinoma cells. *Oncogene*, **10**, 2297–2305.

Jäättelä, M., Mouritzen, H., Elling, F. and Bastholm, L. (1996) A20 zinc finger protein inhibits TNF and IL-1 signalling. *J. Immunol.*, **156**, 1166–1173.

Jackson, M. B. (1990) Hormones and developmental change in plants subjected to submergence or soil waterlogging. *Aquatic Bot.*, **38**, 49–72.

Jazwinski, S. M. (1992) Genes of youth: genetics of ageing in Baker's yeast. *ASM News*, 172–178.

Jeannin, P., Delneste, Y., Seveso, M. *et al.* (1996) IL-12 synergizes with IL-2 and other stimuli in inducing IL-10 production in human T cells. *J. Immunol.*, **156**, 3159–3165.

Jones, G. W. and Bowen, I. D. (1980) The fine structural localization of acid phosphatase in pore cells of embryonic and newly hatched *Deroceras reticulatum* (Pulmonata: Stylommatophora). *Cell Tissue Res.*, **204**, 253–265.

Kagi, D., Ledermann, B., Burki, K. *et al.* (1996) Molecular mechanisms of lymphocyte-mediated cytotoxicity and their role in immunological protection and pathogenetics *in vivo*. *Annu. Rev. Immunol.*, **14**, 207–232.

Kamachi, K., Yamaya, T., Mae, T. and Ojima, K. (1991) A role for glutamine synthetase in the remobilization of leaf nitrogen during natural senescence in rice leaves. *Plant Physiol.*, **96**, 411–417.

Kamens, J., Paskind, M., Hugunin, M. *et al.* (1995) Identification and characterization of ICH-2, a novel member of the interluekin-1β-converting enzyme family of cysteine proteases. *J. Biol. Chem.*, **270**, 15250–15256.

Kerr, J. F. R. (1971) Shrinkage necrosis: a distinct mode of cellular death. *J. Pathol.*, **105**, 13–20.

Kerr, J. F. R., Harmon, B. and Searle, J. (1974) An electron microscope study of cell deletion in the anuran tadpole tail during spontaneous metamorphosis with special reference to apoptosis of striated muscle fibres. *J. Cell Sci.*, **14**, 571–585.

Kerr, J. F. R., Wyllie, A. H. and Currie, A. R. (1972) Apoptosis: a basic biological phenomenon with wide-ranging implications in tissue kinetics. *Br. J. Cancer*, **26**, 239–257.

Khalil, N., Battistuzzi, S. C., Kraut, R. P. *et al.* (1990) Growth factor-initiated proliferation of mouse embryonic fibroblasts induces cytoxicity by natural killer cells and by a non-cytolysin cytotoxin in natural killer granules. *J. Immunol.*, **145**, 1286–1292.

Kiefer, M. C., Brauer, M. J., Powers, V. C. *et al.* (1995) Modulation of apoptosis by the widely distributed Bcl-2 homologue Bak. *Nature*, **374**, 736–739.

Kinzler, K. W. and Vogelstein, B. (1996) Life (and death) in a malignant tumour. *Nature*, **379**, 19–20.

Kischkel, F. C., Hellbardt, S., Behrmann, I. *et al.* (1995) Cytoxicity-dependent APO-1 (Fas/CD95)-associated proteins (CAP) form a death-inducing signalling complex (DISC) with the receptor. *EMBO J.*, **14**, 5579—5588.

Kishimoto, T. (1989) The biology of interleukin-6. *Blood*, **74**, 1–10.

Kishimoto, T., Taga, T. and Akira, S. (1994) Cytokine signal transduction. *Cell*, **76**, 253–262.

Knudson, C., Tunk, K., Tourtellotte, W. *et al.* (1995) Bax-deficient mice with lymphoid hyperplasia and male germ cell death. *Science*, **270**, 96–99.

Kos, F. J. and Engleman, E. G. (1996) Immune regulation: a critical link between NK cells and CTLs. *Immunol Today*, **17**, 174–176.

Kozopas, K. M., Yang, T., Buchan, H. L. *et al.* (1993) MCL-1, a gene expressed in programmed myeloid-cell differentiation, has sequence similarity to BCL-2 *Proc. Natl Acad. Sci. USA*, **90**, 3516–3520.

Krammer, P. H., Behrmann, I., Daniel, P. *et al.* (1994) Regulation of apoptosis in the immune system. *Curr. Opin. Immunol.*, **6**, 279–289.

Kuida, K., Zheng, T. S., Na, S. *et al.* (1996) Decreased apoptosis in the brain and premature lethality in CPP-deficient mice. *Nature*, **384**, 368–384.

Kumagai, A. and Dunphy, W. G. (1991) The cdc25 protein controls tyrosine dephosphorylation of the cdc2 protein in a cell-free system. *Cell*, **64**, 903–914.

Kumar, S., Kinoshita, M., Noda, M. *et al.* (1994) Induction of apoptosis by the mouse *nedd2* gene, which encodes a protein similar to the product of the *Caenorhabditis elegans* cell death gene *ced*-3 and the mammalian IL-1β-converting enzyme. *Genes Dev.*, **8**, 1613–1626.

Lanier, L. L. and Phillips, J. H. (1996) Inhibitory MHC class 1 receptors on NK cells and T cells. *Immunol. Today*, **17**, 86–91.

Lawrence, P. A. (1973) The development of spatial patterns in the integument of insects, in *Developmental Systems: Insects* (eds S. J. Counce and C. H. Waddington), vol. 2, Academic Press, London, pp. 157–209.

Lazebnik, Y. A., Cole, S. Cooke, C. A. *et al.* (1993) Nuclear events of apoptosis in vitro in cell-free mitotic extracts: a model system for analysis of the active phase of apoptosis. *J. Cell Biol.*, **123**, 7–22.

Lazebnik, Y. A., Takahashi, A., Moir, R. D. *et al.* (1995) Studies of the lamin proteinase reveal multiple parallel biochemical pathways during apoptotic execution. *Proc. Natl Acad. Sci. USA*, **92**, 9042–9046.

Lee, M. and Nurse, P. (1988) Cell cycle control genes in fission yeast and mammalian cells. *Trends Genet.*, **4**, 287–290.

Levi-Montacini, R. and Aloe, L. (1981) Mechanisms of action of nerve growth factor in intact and lethally injured sympathetic nerve cells in neonatal rodents, in *Cell Death in Biology and Pathology* (eds I. D. Bowen and R. A. Lockshin), Chapman & Hall, London, pp. 295–327.

164 *References*

Levy, Y. and Bronet, J. C. (1994) Interleukin-10 prevents spontaneous death of germinal center B cells by induction of the bcl-2 protein. *J. Clin. Invest.*, **93**, 424–428.

Li, R., Waga, S., Hannon, G. *et al.* (1994) Differential effects by the p21 cdk inhibitor on PCNA dependent DNA replication and DNA repair. *Nature*, **371**, 534–537.

Lichtenheld, M., Olsen, K. J., Lu, P. *et al.* (1988) Structure and function of human perforin. *Nature*, **335**, 448–451.

Lichtenstein, A., Tu, Y., Fady, C. *et al.* (1995) Interleukin-6 inhibits apoptosis of malignant plasma cells. *Cell Immunol.*, **162**, 248–255.

Lin, E., Orlofsky, A., Berger, M. and Prystowsky, M. (1993) Characterization of A1, a novel hemopoietic-specific early-response gene with sequence similarity to *bcl*-2. *J. Immunol.*, **151**, 1979–1988.

Lin, J., Chen, J., Elenbaas, B. and Levine, A. J. (1994) Several hydrophobic amino acids in the p53 amino-terminal domain are required for transcriptional activation, binding to mdm-2 and the adenovirus E1B 55-kD protein. *Genes Dev.*, **8**, 1235–1246.

Lindahl, T., Satoh, M. S., Poirier, G. G. and Klungland, A. (1995) Post-translational modification of poly (ADP-ribose) polymerase induced by DNA strand breaks. *Trends Biochem. Sci.*, **20**, 405–411.

Linke, S. P., Clarkin, F. C., Di Leonardo, A. *et al.* (1996) A reversible p53 dependent G_0/G_1 cell cycle arrest induced by ribonucleotide depletion in the absence of detectable DNA damage. *Genes Dev.*, **10**, 934–947.

Lippke, J. A., Gu, Y., Sarnecki, C. *et al.* (1996) Identification and characterization of CPP-32/Mch-2 homolog 1, a novel cysteine protease similar to CPP-32. *J. Biol. Chem.*, **271**, 1825–1828.

Liu, C.-C., Walsh, C. M. and Young, J. D.-E. (1995) Perforin: structure and function. *Immunol. Today*, **16**, 194–201.

Liu, C.-C., Steffen, M., King, F. and Young, D. W. (1987) Identification, isolation and characterization of a novel cytotoxin in murine cytolytic lymphocytes. *Cell*, **51**, 393–403.

Liu, X., Kim, C. W., Yang, J. *et al.* (1996) Induction of apoptotic program in cell-free extracts: requirement for dATP and cytochrome c. *Cell*, **86**, 147–157.

Lobe, C. G., Finlay, B. B., Paranchych, W. *et al.* (1986) Novel serine proteases encoded by two cytotoxic T-lymphocyte-specific genes. *Science*, **232**, 858–861.

Lockshin, R. A. (1969) Programmed cell death. Activation of lysis by a mechanism involving the synthesis of protein. *J. Insect Physiol.*, **15**, 1505–1516.

Lockshin, R. A. (1971) Programmed cell death: nature of the nervous signal controlling breakdown of intersegmental muscles. *J. Insect Physiol.*, **17**, 149–158.

Lockshin, R. A. and Williams C. M. (1964) Programmed cell death II. Endocrine potentiation of the breakdown of the intersegmental muscles of silkmoths. *J. Insect Physiol.*, **10**, 643–649.

Lockshin, R. A. and Zakeri, Z. (1996) The biology of cell death and its relationship to ageing, in *Cellular Aging and Cell Death* (eds N. J. Holbrook, G. R. Martin and R. A. Lockshin), Wiley–Liss, New York, pp. 167–180.

Lodish, H., Baltimore, D., Berk, A. *et al.* (1995) *Molecular Cell Biology*, Scientific American Books, W. H. Freeman & Co., New York.

Looss, A. (1889) Über Degenerations Erscheinungen im Tierreich, besonders über die Reduktion des Forschlarvenschwanzes und die in Verlaug disselben äußertenden Histolytischen. *Proz. Preisschr. Jablonowsk Ges.*, **10**, 1–57.

Los, M., Van de Craen, M., Penning, L. C. *et al.* (1995) Requirement of an ICE/CED-3 protease for Fas/APO-1 mediated apoptosis. *Nature*, **375**, 81–83.

Louahead, J., Kermouni, A., Van Snick, J. and Renauld, J. C. (1995) IL-9 induces expression of granzyme and high affinity IgE receptor in murine T-helper clones. *J. Immunol.*, **154**, 5061–5070.

Lowe, S. W. (1996) The role of p53 in apoptosis, in *Apoptosis in Normal Development and Cancer* (ed. M. Slyser), Taylor & Francis, London, pp. 97–125.

Lu, H., Zawel, L., Fisher, L. *et al.* (1992) Human general transcription factor II H phosphorylates the c-terminal domain of RNA polymerase II. *Nature*, **358**, 641–645.

McGowan, C. H., Russell, P. and Reed, S. I. (1990) Periodic biosynthesis of the human M-phase promoting factor catalytic subunit p34 during the cell cycle. *Mol. Cell. Biol.*, **10**, 3847–3851.

Manfredini, R., Grande, A., Tagliafico, E. *et al.* (1993) Inhibition of c-*fes* expression by an antisense oligomer causes apoptosis of HL60 cells induced to granulocytic differentiation. *J. Exp. Med.*, **178**, 381–389.

Mannel, D., Murray, C., Risau, W. and Clauss, M. (1995) Tumor necrosis: factors and principles. *Trends Immunol. Today*, **17**, 254–256.

Margassi, L. and Lawrence, P. A. (1988) The pattern of cell death in *fushi tarazu*, a segmentation gene of *Drosophila. Development*, **104**, 447–451.

Martin, D. P. and Johnson, J. (1991) Programmed cell death in the peripheral nervous system, in *Apoptosis: The Molecular Basis of Cell Death* (eds. D. L. Tomei and F. O. Cope), Current Communications in Cell and Molecular Biology, vol. 3, Cold Spring Harbor Laboratory Press, New York, pp. 247–261.

Martin, D. P., Schmidt, R. E., De Stefano, P. S. *et al.* (1988) Inhibitors of protein synthesis prevent neuronal death caused by nerve growth factor deprivation. *J. Cell Biol.*, **106**, 829–840.

Martin, S. J., O'Brien, G. A., Nishioka, W. K. *et al.* (1995) Proteolysis of fodrin (non-erythroid spectrin) during apoptosis. *J. Biol. Chem.*, **270**, 6425–6428.

Martin, J., Amarante-Mendes, G. P., Shi, L. *et al.* (1996) The cytotoxic cell protease granzyme B initiates apoptosis in a cell free system by proteolytic processing and activation of ICE/CED3 family protease, CPP32, via a novel two-step mechanism. *EMBO J.*, **15**, 2407–2416.

Meijer, L., Ostvold, A., Walaas, S. *et al.* (1991) High mobility group (HMG) proteins I, Y and P1 as substrates of the M phase-specific p34^{cdc}/cyclin^{cdc12} kinase. *Eur. J. Biochem.*, **196**, 557–567.

Milligan, C. E. and Schwartz, L. M. (1996) Programmed cell death during development of animals, in *Cellular Aging and Cell Death*, (eds N. J. Holbrook, G. R. Martin and R. A. Lockshin), Wiley–Liss, New York, pp. 181–208.

Minn, A. J., Velez, P., Schendel, S. L. *et al.* (1997) Bcl-X$_L$ forms an ion channel in synthetic lipid membranes. *Nature*, **385**, 353–357.

Minshall, C., Arkins, S., Freund, G. G. and Kelley, K. W. (1996) Requirement for phosphatidylinositol 3-kinase to protect hemopoietic progenitors against apoptosis depends upon the extracellular survival factor. *J. Immunol.*, **156**, 939–947.

Mittler, R., Shulaev, V. and Lam, E. (1995) Coordinated activation of programmed cell death and defence mechanisms in transgenic tobacco plants expressing a bacterial proton pump. *Plant Cell*, **7**, 29–42.

Miura, M., Zhu, H., Rotello, R. *et al.* (1993) Induction of apoptosis in fibroblasts by IL-1B-converting enzyme, a mammalian homolog of the *C. elegans* cell death gene *ced-3. Cell*, **75**, 653–660.

Miyashita, T., Krajewski, S., Krajewska, M. *et al.* (1994) Tumor suppressor p53 is a regulator of *bcl-2* and *bax* in gene expression *in vitro* and *in vivo. Oncogene*, **9**, 1799–1805.

Moore, K. W., O'Garra, A. O., de Waal Malefyt R. *et al.* (1993) Interleukin-10. *Annu. Rev. Immunol.*, **11**, 165–190.

Mor, F. and Cohen I. R. (1996) IL-2 rescues antigen-specific T-cells from radiation or dexamethasone-induced apoptosis. *J. Immunol.*, **156**, 515–522.

Moreno, S. and Nurse, P. (1994) Regulation of progression through the G_1 phase of the cell cycle by the *rum1+* gene. *Nature*, **367**, 236–242.

Morgan, D. A., Ruscetti, F. W. and Gallo, R. C. (1976) Selective *in vitro* growth of T-lymphocytes from normal human bone marrow. *Science*, **193**, 1007–1008.

Morgenbesser, S. D., Williams, B. O., Jacks, T. and DePinho, R. R. (1994) p53-dependent apoptosis produced by Rb-deficiency in the developing mouse lens. *Nature*, **371**, 72–74.

Moss, M. L., Jin, S.-L. C., Milla, M. E. *et al.* (1997) Cloning of a disintegrin metallo-proteinase that processes precursor tumour-necrosis factor-α. *Nature*, **385**, 733–736.

Mountz, J. D., Wu, J., Cheng, J. and Zhou, T. (1994) Autoimmune disease: a problem of defective apoptosis. *Arthritis Rheum.*, **37**, 1415–1420.

Munday, N. A., Vaillancourt, J. P., Ali, A. *et al.* (1995) Molecular cloning and proapoptotic activity of ICE rel-11 and ICE rel-111, members of the ICE/CED-3 family of cysteine proteases. *J. Biol. Chem.*, **270**, 15870–15876.

Musacchio, A., Gibson, T., Rise, P. *et al.* (1993) The PH domain – a common piece in the structural patchwork of signaling proteins. *Trends Biochem. Sci.*, **18**, 343–348.

Muzio, M., Chinnaiyan, A. M., Kiskchkel, F. C. *et al.* (1996) Flice, a novel FADD-homologous ICE/CED-3 like protease, is recruited to the CD95 (Fas/APO-1) death-inducing signaling complex. *Cell*, **85**, 817–827.

Nagafuji, K., Shibuya, T., Harada, M. *et al.* (1995) Functional expression of Fas antigen (CD95) on hematopoietic progenitor cells. *Blood*, **86**, 883–889.

Nagata, S. and Goldstein, P. (1995) The Fas death factor. *Science*, **267**, 1449–1456.

Nakata, M., Smyth, M. J., Norishisa, Y. *et al.* (1990) Constitutive expression of pore forming protein in peripheral blood γ/δ T-cells: implications for their cytotoxic role in vivo. *J. Exp. Med.*, **172**, 1877–1880.

Nasmyth, K. (1993) Control of the yeast cell cycle by the cdc28 protein kinase. *Curr. Opin. Cell Biol.*, **5**, 166–179.

Neilan, J., Lu, Z., Afonso, C. *et al.* (1993) An African swine fever virus gene with similarity to the proto-oncogene *bcl-2* and the Epstein–Barr virus gene, *BHRF1*. *J. Virol.*, **67**, 4391–4394.

Nevins, J. (1992) E2F: a link between the Rb tumour suppressor protein and viral oncoproteins. *Science*, **258**, 424–429.

Nicholson, D. W., Ali, A., Thornberry, N. A. *et al.* (1995) Identification and inhibition of the ICE/CED-3 protease necessary for mammalian apoptosis. *Nature*, **376**, 37–43.

Nishioka, W. K. and Welsh, R. M. (1994) Susceptibility to cytotoxic T lymphocyte-induced apoptosis is a function of the proliferative status of the target. *J. Exp. Med.*, **179**, 769–774.

Nurse, P. (1990) Universal control mechanisms regulating onset of M phase. *Nature*, **344**, 503–508.

Nurse, P. (1994) Ordering of S phase and M phase in the cell cycle. *Cell*, **79**, 547–550.

Oberhammer, F., Frisch, G., Schmied, M. *et al.* (1993) Condensation of the chromatin at the membrane of an apoptotic nucleus is not associated with activation of an endonuclease. *J. Cell Sci.*, **104**, 317–326.

Ohtsubo, M., Theodres, A., Schumacher, J. *et al.* (1995) Human cyclin E, a nuclear protein essential for the G_1 to S phase transition. *Mol. Cell Biol.*, **15**, 2612–2624.

Onishi, Y., Azuma, Y., Sato, Y. *et al.* (1993) Topoisomerase inhibitors induce apoptosis in thymocytes. *Biochim. Biophys. Acta*, **1175**, 147–154.

Opipari, A. W., Hu, H. M., Yabkowitz, R. and Dixit, V. M. (1992) The A20 zinc-finger protein protects cells from TNF cytotoxicity. *J. Biol. Chem.*, **267**, 12424–12427.

Owen-Schaub, L. B., Zhang, W., Cusack, J. C. *et al.* (1995) Wild-type human p53 and a temperature-sensitive mutant induce Fas/APO-1 expression. *Mol. Cell Biol.*, **15**, 3032–3040.

Owens, G. P., Hahn, W. E. and Cohen, J. J. (1991) Identification of mRNAs associated with programmed cell death in immature thymocytes. *Mol. Cell Biol.*, **11**, 4177–4188.

Oyaizu, N., McCloskey, T. W., Than, S. *et al.* (1994) Cross-linking of CD4 molecule upregulates Fas antigen expression in lymphocytes by inducing interferon-γ and tumour necrosis factor-α secretion. *Blood*, **84**, 2622–2631.

Pawson, T. (1995) Protein modules and signalling networks. *Nature*, **373**, 573–580.

Pearson, G. R., Luka, J., Petti, L. *et al.* (1987) Identification of an Epstein–Barr virus early gene encoding a second component of the restricted early antigen complex. *Virology*, **160**, 151–161.

Perez-Terzic, C., Pyle, S., Jaconi, M. *et al.* (1996) Conformational states of the nuclear pore complex induced by depletion of nuclear Ca^{2+} stores. *Science*, **273**, 1875–1877.

Peter, M. and Heiskowitz, I. (1994) Joining the complex: cyclin-dependent kinase inhibitory proteins and the cell cycle. *Cell*, **79**, 181–184.

Peter, M. E., Kischkel, F. C., Hellbardt, S. *et al.* (1996) CD95(APO-1/Fas)-associating signaling proteins. *Cell Death Differ.*, **3**, 161–170.

Pinkoski, M. J., Winkler, U., Hudig, D. and Bleackley, R. C. (1996) Binding of granzyme B in the nucleus of target cells. Recognition of an 80-beta delta protein. *J. Biol. Chem.*, **271**, 10225–10229.

Polyak, K., Xia, Y.,, Zweier, K. W. *et al.* (1997) A model for p53-induced apoptosis. *Nature*, **389**, 300–305.

Poon, R. and Hunt, T. (1992) Identification of the domains in cyclin A required for binding to, and activation of, $p34^{cdc2}$ and $p32^{cdk2}$ protein kinase subunits. *Mol. Biol. Cell*, **3**, 1279–1294.

Potten, C. S. (1992) The significance of spontaneous and induced apoptosis in the gastrointestinal tract. *Cancer Metastasis Rev.*, **11**, 179–195.

Prestige, M. C. (1970) Differentiation, degeneration and the role of the periphery: quantitative considerations, in *The Neurosciences* (ed. F. O. Schmitt), Rockefeller University Press, New York, pp. 73–82.

Quan, L. T., Caputo, A., Bleackley, R. C. *et al.* (1995) Granzyme B is inhibited by the cowpox virus serpin cytokine response modifier A. *J. Biol. Chem.*, **270**, 10377–10379.

Quan, L. T., Tewari, M., O'Rourke K. *et al.* (1996) Proteolytic activation of the cell death protease/yama/CPP32 by granzymeB. *Proc. Natl Acad. Sci. USA*, **93**, 1972–1976.

Raff, M. C., Barres, B. A., Burne, J. F. *et al.* (1994) Programmed cell death and the control of cell survival. *Phil. Trans. R. Soc. Lond. B*, **345**, 265–268.

Ray, C. A., Black, R. A., Kronheim, S. R. *et al.* (1992) Viral inhibition of inflammation: cowpox virus encodes an inhibitor of the interleukin-1β-converting enzyme. *Cell*, **69**, 597–604.

Redelman, D. and Hudig, D. (1980) The mechanism of cell-mediated cytotoxicity. I. Killing by murine cytotoxic T lymphocytes requires cell surface thiols and activated proteases. *J. Immunol.*, **124**, 870–887.

Reed, J. C. (1996) Bcl-2 and the regulation of programmed cell death in cancer, in

Apoptosis in Normal Development and Cancer (ed. M. Slyser), Taylor & Francis, London, pp. 127–159.

Reed, J. C. (1997) Double identity for proteins of the Bcl-2 family. *Nature*, **387**, 773–776.

Rensing-Ehl, A., Hess, S., Ziegler-Heitbrock, W. *et al.* (1995) Fas/Apo-1 activates nuclear factor kappa B and induces interleukin-6 production. *J. Inflamm.*, **45**, 161–174.

Rubin, L. L., Philpott, K. L and Brooks, S. F. (1993) The cell cycle and cell death. *Curr. Biol.*, **3**, 390–394.

Sabbatini, P., Chiou, S. K., Rao, L. and White, E. (1995) Modulation of p53-mediated transcription and apoptosis by adenovirus E1B 19K protein. *Mol. Cell Biol.*, **15**, 1060–1070.

Salcedo, T. W., Azzoni, L., Wolf, S. F. and Perussia, B. (1993) Modulation of perforin and granzyme messenger RNA expression in human natural killer cells. *J. Immunol.*, **151**, 2511–2520.

Sarraf, C. E. and Bowen, I. D. (1986) Kinetic studies on a murine sarcoma and an analysis of apoptosis. *Br. J. Cancer*, **54**, 989–998.

Sarraf, C. E. and Bowen, I. D. (1988) Proportions of mitotic and apoptotic cells in a range of untreated experimental tumours. *Cell Tissue Kinet.*, **21**, 45–49.

Sato, T., Hanada, M., Bodrug, S. *et al.* (1994) Interactions among members of the bcl-2 protein family analysed with a yeast two-hybrid system. *Proc. Natl Acad. Sci. USA*, **91**, 9238–9242.

Saunders, J. W. (1966) Death in embryonic systems. *Science*, **154**, 604–612.

Saunders, J. W. Jr, Gasseling, M. T. and Saunders, L. C. (1962) Cellular death in morphogenesis of the avian wing. *Dev. Biol.*, **5**, 147–178.

Savill, J., Fadok, V., Henson, P. and Haslett, C. (1993) Phagocyte recognition of cells undergoing apoptosis. *Immunol. Today*, **14**, 131–136.

Schindler, T., Bergfield, R. and Schopfer, P. (1995) Arabino-galactan proteins in maize coleoptiles: developmental relationship to cell death during xylem differentiation but not extension growth. *Plant J.*, **7**, 25–36.

Schlegel, J., Peter, I., Orrenius, S. *et al.* (1996) CPP-32/apopain is the key interleukin-1β-converting enzyme-like protease involved in Fas-mediated apoptosis. *J. Biol. Chem.*, **271**, 1841–1844.

Schwartz, L. M. (1991) The role of cell death genes during development. *BioEssays*, **13**, 389–395.

Schwartz, L. M., Kosz, L. and Kay, B. K. (1990) Gene activation is required for developmentally programmed cell death. *Proc. Natl Acad. Sci. USA*, **87**, 6594–6599.

Schwartz, L. M. and Osborne, B. A. (1993) Programmed cell death, apoptosis and killer genes. *Immunol. Today*, **14**, 582–590.

Schwarze, M. M. and Huwley, R. G. (1995) Prevention of myeloma cell apoptosis by ectopic bcl-2 expression or interleukin-6 mediated up-regulation of bcl-X_L. *Cancer Res.*, **55**, 2262–2265.

Schweichel, J. U. and Merker, H. J. (1973) The morphology of various types of cell death in pre-natal tissues. *Teratology*, **7**, 253–266.

Selleri, C., Sato, T., Anderson, S. *et al.* (1995) Interferon-γ and tumour necrosis factor-α suppress both early and late stages of hematopoiesis and induce programmed cell death. *J. Cell Physiol.*, **165**, 538–546.

Server, A. C and Mobley, W. C. (1991) Neuronal cell death and the role of apoptosis, in *Apoptosis: The Molecular Basis of Cell Death* (eds. D. L. Tomei and F. O. Cope), Current Communications in Cell and Molecular Biology, vol. 3, Cold Spring Harbor Laboratory Press, New York, pp. 263–278.

Shen, Y. and Shenk, T. (1994) Relief of p53 mediated transcriptional repression by the adenovirus E1B 19-kDa protein in the cellular Bcl-2 protein. *Proc. Natl Acad. Sci. USA*, **91**, 8940–8944.

Sherwood, S. W. and Schimke, R. T. (1994) Induction of apoptosis by cell-cycle phase specific drugs, in *Apoptosis* (eds E. Mihich and R. T. Schimke), Plenum Press, New York, pp. 223–236.

Shi, L., Krant, R. P., Aebersold, R. and Greenberg, A. H. (1992) A natural killer cell granule protein that induces DNA fragmentation and apoptosis. *J. Exp. Med.*, **175**, 553–566.

Shi, L., Nishioka, W. K., Th'ng, J. *et al.* (1994) Premature p34^{cdc2} activation required for apoptosis. *Science*, **263**, 1143–1145.

Shi, L., Chen, G., MacDonald, G. *et al.* (1996a) Activation of an interleukin 1 converting enzyme-dependent apoptosis pathway by granzyme B. *Proc. Natl Acad. Sci. USA*, **93**, 11002–11007.

Shi, L., Chen, G., He, D. *et al.* (1996b) Granzyme B induces apoptosis and cyclin A-associated cyclin dependent kinase activity in all stages of the cell cycle. *J. Immunol.*, **157**, 2381–2385.

Shih, S. C. and Stutman, O. (1996) Cell cycle-dependent tumour necrosis factor and apoptosis. *Cancer Res.*, **56**, 1591–1598.

Shinkai, Y., Takio, K. and Okumura, K. (1988) Homology of perforin to the ninth component of complement (C9). *Nature*, **334**, 525–527.

Sigee, D. C. (1984) Induction of leaf cell death by phytopathogenic bacteria, in *Cell Ageing and Cell Death* (eds I. Davies and D. C. Sigee), S. E. B. Seminar Series 25, Cambridge University Press, Cambridge, pp. 295–322.

Silvennoinen, O., Schindler, C., Schlessinger, J. and Levy, D. E. (1993) Ras-independent growth factor signaling by transcription factor tyrosine phosphorylation. *Science*, **261**, 1736–1739.

Singleton, J. R., Dixit, V. M. and Feldman, E. L. (1996) Type I insulin-like growth factor receptor activation regulates apoptotic proteins. *J. Am. Soc. Biochem. Mol. Biol.*, **271**, 31791–31794.

Smart, C. M. (1994) Gene expression during leaf senescence, Tansley Review 63. *New Phytol.*, **126**, 419–448.

Smith, C. A., Williams, G. T., Kingston, R. *et al.* (1989) Antibodies to CD3/T-cell receptor complex induce death by apoptosis in immature T-cells in thymic cultures. *Nature*, **337**, 181–184.

Smyth, J. and Trapani, J. A. (1995) Granzyme: exogenous proteinases that induce target cell apoptosis. *Immunol. Today*, **16**, 202–204.

Smyth, M. J., Strobl, S. L., Young, H. A. *et al.* (1991) Regulation of lymphokine-activated killer activity and pore-forming protein gene expression in human peripheral blood CD8 T lymphocytes. *J. Immunol.*, **146**, 3289–3297.

Soussi, T. (1996) The humoral response to the tumor-suppressor gene-product p53 in human cancer: implications for diagnosis and therapy. *Immunol. Today*, **17**, 354–356.

Spooner, E., Fairburn, L. J., Cowling, G. J. *et al.* (1994) Biological consequences of p160^{v-abl} protein tyrosine kinase activity in a primitive multipotent haemopoietic cell line. *Leukemia*, **8**, 620–630.

Srinivasula, S. M., Fernandes-Alnemri, T., Zangrilli, J. *et al.* (1996) The Ced-3/interleukin 1β converting enzyme-like homolog Mch6 and the lamin-cleaving enzyme Mch2α are substrates for the apoptotic mediator CPP32. *J. Biol. Chem.*, **271**, 27099–27106.

Steel, G. G. (1966) Delayed uptake by tumours of tritium from thymidine. *Nature*, **210**, 806–808.

Steel, G. G. (1968) Cell loss from experimental tumours. *Cell Tissue Kinet.*, **1**, 193–207.

Steel, G. G. (1977) *Growth Kinetics of Tumours*, Oxford University Press, Oxford.

Strasser, A., Harris, A. W., Jack, T. and Cory, S. (1994) DNA damage can induce apoptosis in proliferating lymphoid cells via p53 independent mechanisms inhibitable by *bcl-2*. *Cell*, **79**, 329–339.

Strausfield, V., Labbe, J. C., Fesquat, O. *et al.* (1991) Dephosphorylation and activation of a p34^{cdc2}/cyclin B complex *in vitro* by human cdc25 protein. *Nature*, **35**, 242–245.

Su, B. and Karin, M. (1996) Mitogen-activated protein kinase cascades and regulation of gene expression. *Curr. Opin. Immunol.*, **8**, 402–411.

Suda, T. and Nagata, S. (1994) Purification and characterisation of the Fas-ligand that induces apoptosis. *J. Exp. Med.*, **179**, 873–879.

Suda, T., Takahashi, T., Golstein, P. and Nagata, S. (1993) Molecular cloning and expression of the fas ligand, a novel member of the tumor necrosis factor family. *Cell*, **75**, 1169–1178.

Sugimoto, A., Freisen, P. D. and Rothman, J. H. (1994) Baculovirus p35 prevents developmentally programmed cell death and rescues a *ced-9* mutant in the nematode *Caenorhabditis elegans*. *EMBO J.*, **13**, 2023–2028.

Szegedi, Z., Stetak, A., Nagy, J. et al. (1996) Characterisation of cell death caused by tyrphostin AG-213 on human cell lines. *Cell Prolif.*, **29**, 6.

Szende, B., Keri, G., Szegedi, Z. *et al.* (1995) Tyriphostin induces non-apoptotic programmed cell death in colon tumour cells. *Cell Biol. Int.*, **19**, 903–911.

Takayama, S., Sato, T., Krajewski, S. *et al.* (1995) Cloning and functional analysis of Bag-1: a novel Bcl-2 binding protein with anti-cell death activity. *Cell*, **80**, 279–284.

Tannenbaum, C. S., Wicker, N., Armstrong, D. *et al.* (1996) Cytokine and chemokine expression in tumors of mice receiving systemic therapy with IL-12. *J. Immunol.*, **156**, 693–699.

Tewari, M. and Dixit, V. M. (1995) Fas-b and tumour necrosis factor-induced apoptosis is inhibited by pox-virus *crm* A gene product. *J. Biol. Chem.*, **270**, 3255–3260.

Tewari, M., Quan, L. T., O'Rourke, K. *et al.* (1995) Yama/CPP-32β, a mammalian homolog of CED-3, is a CrmA-inhibitable protease that cleaves the death substrate poly(ADP-ribose) polymerase. *Cell*, **81**, 801–809.

Thomas, H. (1987) *Sid*: a Mendelian locus controlling thylakoid membrane disassembly in senescing leaves of *Festuca pratensis*. *Theor. Appl. Genet.*, **73**, 551–555.

Thomas, H. (1994) Ageing in the plant and animal kingdoms – the role of cell death. *Rev. Clin. Gerontol.*, **4**, 5–20.

Thomas, H., Ougham, H. J. and Davies, T. G. E. (1992) Leaf senescence in a non-yellowing mutant of *Festuca pratensis*: transcripts and translation products. *J. Plant Physiol.*, **139**, 403–412.

Thome, M., Schneider, P., Hofmann, K. *et al.* (1997) Viral FLICE-inhibitory proteins (FLIPS) prevent apoptosis induced by death receptors. *Nature*, **386**, 517–521.

Thornberry, N. A., Bull, H. G., Calaycay, J. R. *et al.* (1992) A novel heterodimeric cysteine protease is required for interleukin-1β processing in monocytes. *Nature*, **356**, 768–744.

Tian, Q., Streuli, M., Saito, H. *et al.* (1991) A polyadenylate binding protein localised to the granules of cytolytic lymphocytes induces DNA fragmentation in target cells. *Cell*, **67**, 629–639.

Trapani, J. A., Browne, K. A., Smyth, M. J. and Jans, D. A. (1996) Localization of granzyme B in the nucleus: a putative role in the mechanism of cytotoxic lymphocyte-mediated apoptosis. *J. Biol. Chem.*, **271**, 4127–4133.

Trump, B. F., Berezesky, I. K. and Osorino-Vargas, A. R. (1981) Cell death and the disease process: the role of calcium, in *Cell Death in Biology and Pathology* (eds I. D. Bowen and R. A. Lockshin), Chapman & Hall, London, pp. 209–242.

Tsuboi, M., Eguchi, K., Kawakami, A. *et al.* (1996) Fas antigen expression on synovial cells was down-regulated by interleukin-1-β, *Biochem. Biophys. Res. Commun.*, **218**, 280–285.

Ucker, D. S. (1987) Cytotoxic T-lymphocytes and glucocorticoids activate an endogenous suicide process in target cells. *Nature*, **327**, 62–64.

Ucker, D. S. (1991) Death by suicide: one way to go in mammalian development? *N. Biol.*, **3**, 103–109.

Ucker, D. S., Obermiller, P. S., Eckart, W. *et al.* (1992) Genome digestion is a dispensable consequence of physiological cell death mediated by cytotoxic T lymphocytes. *Mol. Cell Biol.*, **12**, 3060–3069.

Ullrich, A. and Schlessinger, J. (1990) Signal transduction by receptors with tyrosine kinase activity. *Cell*, **61**, 203–212.

Urban, J. L., Shepard, H. M., Rothstein, J. L. *et al.* (1986) Tumour necrosis factor: a potent effector molecule for tumour cell killing by activated macrophage. *Proc. Natl Acad. Sci. USA*, **83**, 5233–5237.

Vaux, D. L. (1996) Evolution of apoptosis, in *Apoptosis in Normal Development and Cancer* (ed. M. Slyser), Taylor & Francis, London, pp. 189–200.

Verheij, M., Bose, R., Lin, X. H. *et al.* (1996) Requirement for ceramide-initiated SAPK/JNK signalling in stress induced apoptosis. *Nature*, **380**, 75–79.

Vujanovic, N. L., Nagashima, S., Herberman, R. B. and Whiteside, T. L. (1996) Non-secretory apoptotic killing by human NK cells. *J. Immunol.*, **157**, 1117–1126.

Walsh, C. M., Glass, A. A., Chiu, U. and Clark, W. R. (1994) The role of the Fas lytic pathway in a perforin-less CTL hybridoma. *J. Immunol.*, **153**, 2506–2514.

Wang, H. G., Rapp, U. R. and Reed, J. C. (1996) Bcl-2 targets the protein kinase Raf-1 to mitochondria. *Cell*, **87**, 629–638.

Wang, X. and Ron, D. (1996) Stress induced phosphorylation and activation of the transcription factor CHOP (GADD153) by p38 MAP kinase. *Science*, **272**, 1347–1349.

Wang, L., Miura, M., Bergeron, L. *et al.* (1994) Ich-1 an ICE/ced-3-related gene, encodes both positive and negative regulators of programmed cell death. *Cell*, **78**, 739–750.

Wang, H. G., Takayama, S., Rapp, U. and Reed, J. (1996) Bcl-2 interacting protein BAG-1 binds to and activates the kinase Raf-1. *Proc. Natl Acad. Sci. USA*, **93**, 7063–7068.

Wangenheim, K.-H. (1987) Cell death through differentiation: potential immortality of somatic cells: a failure in control of differentiation, in *Perspectives on Mammalian Cell Death* (ed. C. S. Potten), Oxford University Press, Oxford, ch. 6, pp. 129–159.

Waterhouse, N., Kumar, S., Song, Q. *et al.* (1996) Heteronuclear ribonucleoproteins C1 and C2 components of the spliceosome are specific targets of interleukin 1β-converting enzyme-like proteases in apoptosis. *J. Biol. Chem.*, **271**, 29335–29341.

Weber, R. (1969) Tissue involution and lysosomal enzymes during anuran metamorphosis, in *Lysosomes in Biology and Pathology*, vol. 2 (eds J. T. Dingle and H. B. Fell), North Holland, Amsterdam, pp. 437–461.

Wertz, I. E. and Hanley, M. R. (1996) Diverse molecular provocation of programmed cell death. *Trends Biochem. Sci.*, **21**, 359–364.

White, E. (1994) p53, guardian of Rb. *Nature*, **371**, 21–22.

White, E. (1996) Life, death and the pursuit of apoptosis. *Genes Dev.*, **10**, 1–15.

White, K., Grether, M. E., Abrams, J. M. *et al.* (1994) Genetic control of programmed cell death in *Drosophila*. *Science*, **264**, 677–683.

Wolff, T. and Ready, D. F. (1991) Cell death in normal and rough eye mutants in *Drosophila*. *Development*, **113**, 825.

Woolhouse, H. W. (1984) Senescence in plant cells, in *Cell Ageing and Cell Death* (eds I. Davies and D. C. Sigee), S. E. B. Seminar Series 25, Cambridge University Press, Cambridge, pp. 123–153.

Wright, N. A. (1981) The tissue kinetics of cell loss, in *Cell Death in Biology and Pathology* (eds I. D. Bowen and R. A. Lockshin), Chapman & Hall, London, pp. 171–207.

Wyllie, A. H. (1980) Glucocorticoid-induced thymocyte apoptosis is associated with endogenous endonuclease activation. *Nature*, **284**, 555–556.

Wyllie, A. H. (1981) Cell death: a new classification separating apoptosis from necrosis, in *Cell Death in Biology and Pathology* (eds I. D. Bowen and R. A. Lockshin), Chapman & Hall, London, ch. 1, pp. 9–34.

Wyllie, A. H. (1987) Apoptosis: cell death in tissue regulation. *J. Pathol.*, **153**, 313–316.

Wyllie, A. H. (1997) Clues in the p53 murder mystery. *Nature*, **387**, 237–238.

Wyllie, A. H., Duvall, E. and Blow, J. J. (1984) Intracellular mechanisms in cell death in normal and pathological tissues, in *Cell Ageing and Cell Death* (eds I. Davies and D. C. Sigee), S. E. B. Seminar Series 25, Cambridge University Press, Cambridge, pp. 269–284.

Wyllie, A. H., Morris, R. G., Smith, A. L. and Dunlop, D. (1984) Chromatin cleavage in apoptosis: association with condensed chromatin morphology and dependence in macromolecular synthesis. *J. Pathol.*, **142**, 67–77.

Yamakita, Y., Yamashiro, S. and Matsumura, F. (1992) Characterization of mitotically phosphorylated caldesmon. *J. Biol. Chem.*, **67**, 12022–12029.

Yang, E., Zha, J., Jockel, J. *et al.* (1995) Bad, a heterodimeric partner for Bcl-X$_L$ and Bcl-2 ,displaces Bax and promotes cell death. *Cell*, **80**, 285–291.

Yanuck, M., Carbone, D. P., Pendelton, C. D. *et al.* (1993) A mutant p53 tumour suppressor protein is a target for peptide induced CD8[+] cytotoxic T-cells. *Cancer Res.*, **53**, 3257–3261.

Yao, R. and Cooper, G. M. (1995) Requirement for phosphatidylinositol-3-kinase in the prevention of apoptosis by nerve growth factor. *Science*, **267**, 2003–2006.

Yin, X.-M., Oltvai, Z. and Korsemeyer, S. (1994) BH1 and BH2 domains of Bcl-2 are required for inhibition of apoptosis and heterodimerization with bax. *Nature*, **369**, 321–323.

Yin, C., Knudson, M., Korsmeyer, S. J. and Van Dyke, T.(1997) Bax suppresses tumorigenesis and stimulates apoptosis *in vivo*. *Nature*, **385**, 637–640.

Yonish-Rouach, E. (1996) The *p53* tumour suppressor gene: a mediator of G$_1$ growth arrest and of apoptosis. *Experientia*, **52**, 1001–1005.

Yonish-Rouach, E., Resnitzky, D., Lotem, J. *et al.* (1991) Wild-type *p53* induces apoptosis of myeloid leukaemic cells that are inhibited by interleukin-6. *Nature*, **352**, 345–347.

Yu, C.-R., Lin, J.-X., Fink, D. W. *et al.* (1996) Differential utilization of Janus kinase-signal transducer and activator of transcriptor signalling pathways in the stimulation of human natural killer cells by IL-2, IL-12 and INFα. *J. Immunol.*, **157**, 126–137.

Yuan, J. and Horovitz, H. R. (1990) The *Caenorhabditis elegans* genes *ced-3* and *ced-4* act autonomously to cause programmed cell death. *Dev. Biol.*, **138**, 33–41.

Yuan, J., Shaham, S., Ledoux, S. *et al.* (1993) The *C. elegans* cell death gene *ced-3* encodes a protein similar to mammalian interluekin-1β-converting enzyme. *Cell*, **75**, 641–652.

Zha, H., Aimé-Sempé, C., Sato, T. and Reed, J. (1996a) Pro-apoptotic protein Bax heterodimerizes with Bcl-2 and homodimerizes with Bax via a novel domain (BH3) distant from BH1 and BH2. *J. Biol. Chem*, **271**, 7440–7444.

Zha, J., Harada, H., Yang, E. *et al.* (1996b) Serine phosphorylation of death agonist BAD in response to survival factor results in binding to 14-3-3, not Bcl-X$_L$. *Cell*, **87**, 619–628.

Zhang, H., Hannon, G. and Beach, D. (1994) p21-containing cyclin kinases exist in both active and inactive states. *Genes Dev.*, **8**, 1750–1758.

Zhang, X., Giangreco, L., Broome, H. E. *et al.* (1995) Control of CD4 effector fate: transforming growth factor β1 and interleukin 2 synergize to prevent apoptosis and promote effector expansion. *J. Exp. Med.*, **182**, 699–709.

Zheng-gang, L., Hailing, H., Goeddel, D. V. and Karin, M. (1996) Dissection of TNF receptor 1 effector functions: JNK activation is not linked to apoptosis while NF-κB activation prevents cell death. *Cell*, **87**, 565–576.

Zou, H., Henzel, W. J., Liu, X. *et al.* (1997) Apaf-1, a human protein homologous to *C. elegans* CED-4, participates in cytochrome c-dependent activation of caspase-3. *Cell*, **90**, 405–413.

Index

Numbers in *italics* refer to tables; numbers in **bold** refer to illustrations